D1188454

WHAT DID THE ROMANS KNOW?

WHAT DID THE ROMANS KNOW?
AN INQUIRY INTO SCIENCE AND WORLDMAKING

DARYN LEHOUX

THE UNIVERSITY OF CHICAGO PRESS
CHICAGO AND LONDON

DARYN LEHOUX is professor of classics at Queen's University, Kingston, Ontario.

The University of Chicago Press, Chicago 60637
The University of Chicago Press, Ltd., London

Printed in the United States of America
21 20 19 18 17 16 15 14 13 12   1 2 3 4 5

ISBN-13: 978-0-226-47114-3 (cloth)
ISBN-10: 0-226-47114-4 (cloth)

Library of Congress Cataloging-in-Publication Data

Lehoux, Daryn, 1968–
What did the Romans know? : an inquiry into science and worldmaking / Daryn Lehoux.
p. cm.
Includes bibliographical references and index.
ISBN-13: 978-0-226-47114-3 (cloth : alkaline paper)
ISBN-10: 0-226-47114-4 (cloth : alkaline paper)
1. Science, Ancient. 2. Science—History. I. Title.
Q124.95.L44 2012
930.1—dc23
2011029349

∞ This paper meets the requirements of ANSI/NISO Z39.48-1992 (Permanence of Paper).

This book is dedicated to my daughters Zoë and Mari,
one of whom, at the age of six, asked if there were any magnets
in the house "that we don't need anymore."
She'd heard me talking with colleagues and
wanted to try something . . .

CONTENTS

ACKNOWLEDGMENTS

This project began in a startling moment in an undergraduate classroom in 2001, when I first met garlic and magnets, and has slowly grown to its present length as the full implications of that phenomenon emerged. Along the way, a great many people offered helpful suggestions on draft chapters in one form or another, and it is a real pleasure to thank them here. My gratitude goes out to Mary Beagon, Marco Beretta, Alexander Bird, Christián Carman, Anjan Chakravartty, James Collins, Serafina Cuomo, Nick Denyer, Erna Fiorentini, Sophie Gibson, Yves Gingras, Danny Goldstick, Thomas Habinek, Ian Hacking, Harry Hine, Kinch Hoekstra, Brad Inwood, Alexander Jones, Joshua Katz, Philip Kitcher, Jean-Louis Labarrière, Helen Lang, David Langslow, Thomas Laqueur, Geoffrey Lloyd, Kevin McNamee, Stephen Menn, Erica Milam, Gregg Mitman, Tania Munz, Tim Parkin, Lucia Pasetti, Larry Principe, Brendan Quine, David Sedley, Skúli Sigurdsson, Sergio Sismondo, Kyle Stanford, Thomas Sturm, Fernando Vidal, Robert Wardy, Morton White, and Greg Wolff. If I have forgotten anyone, I offer my sincerest apologies.

I would especially like to thank Heinrich von Staden and Lorraine Daston for their support of the project, and for their careful and engaged discussion of many of its chapters. I am grateful to Jay Foster and Michael Gordin, who were especially careful readers, both of whom offered comments on virtually every chapter of the book in one form or another (and sometimes in multiple forms). This book simply could not be what it is without the very kind help of these four individuals. Finally, I would like to express my thanks to Karen Darling and the anonymous readers from the University of Chicago Press, whose suggestions were so helpful in shaping the final manuscript.

I would also like to thank the Institute for Advanced Study, Princeton;

the Max-Planck-Institut für Wissenschaftsgeschichte; the Loeb Classical Library Foundation, Harvard University; and the Social Sciences and Humanities Research Council of Canada for their generous financial support at various stages of this project, and the University of King's College, Halifax; the University of Manchester; and Queen's University for the granting of academic leaves in which to complete the project.

Many of the chapters benefited greatly from discussion at a number of venues, and for stimulating conversation I thank the people at Oxford, St. Andrews, McGill, Manchester, Bologna, the Max Planck Institute, the Institute for Advanced Study, Princeton, the University of Southern California, Caltech, Toronto, Cambridge, Brock, York University, Bristol, and the University of Pennsylvania.

Finally, I thank my wife, Jill Bryant, for her (seemingly unending) support, and for putting up with all the moving.

CHAPTER ONE

The Web of Knowledge

To paraphrase an old saw: *What's in a world?*

Atoms, aprons, artichokes, and aardvarks: *everything.* But in another sense, no isolated things: atoms are divided and combined into smaller and bigger entities at one and the same time. Artichokes evolve in ecosystems.[1] Everything is particulate, everything part of bigger wholes, and change is everywhere all the time. Somewhere in that tangle we find ourselves observing and putting it all together, *making sense.*

We could have done it differently. Indeed, on one way of looking at it, the history of the sciences is virtually a catalogue of different ways of doing it—not all of them successful in the end (but then we should ask what "successful" means and how the qualification "in the end" matters). In this book, I propose an argument that science (a term which I acknowledge to be a loaded one when speaking about antiquity) is best understood to be happening, as the Romans would say, *in medias res*: in the middle of things. Facts make sense—indeed, may only crystallize as facts—within a very large web of . . . of what, exactly? Of preexisting knowledge about the world, to be sure; but facts also situate themselves within a far-reaching social and cultural milieu, not to mention the many interrelationships facts have with our firsthand experiences of the world (which also implicates our basic apparatus for having those experiences: our perceptual and cognitive systems), and we should not forget the important grounding that facts have in the overarching philosophical, mathematical, and/or logical background against which standards of accuracy, truth, and acceptability are framed. Here the

1. Since artichokes are the subjects of selective breeding, we might better say that they evolve in a combination of markets and ecosystems. For a mapping of this hybrid terrain, see Pollan 2001.

magnitude of the problem threatens to overwhelm. Like Jonathan Swift's fleas, the contexts of science have contexts that have contexts that have contexts.

So how can we situate facts in any finite way? Clearly we cannot map out every last detail of every last connection, of every idea, of every last bit of observational evidence, of every logical or ontological framework, that allow a body of facts to make sense. We can, though, have a close look at the most significant of those interconnections to see how they work and what they imply about how we understand the world around us.

In this book I propose to use the historical study of a very remote period, that of Rome from the first century BC to the second century AD, as a focal field to pick up on three prominent threads in recent debates: these are what we might most loosely call the historical, intellectual, and experiential contexts of fact-making. Under these very broad headings, we will look at (among other things) the ethical, political, cultural, and educational contexts of the sciences at Rome and most particularly their interrelationships with intellectual factors (cognitive, comparative, taxonomic), in order to show how these shape and are shaped by experiences of the world. I ask an old-fashioned question about how we understand observation. But instead of drawing the old-fashioned hard-and-fast line between something called "observation" and something called "theory," I argue that we need to look at the domains in which observation is being situated, understood, and processed, because that is where the world we perceive gets put together as a coherent whole.

This book accordingly focuses on the twin strands of how facts come to be, and where they stand in relation to the larger world in which they find themselves. Because of its sometimes rather extreme foreignness, ancient science in general promises to be a fruitful ground in which to examine these foundational questions in the history and philosophy of the sciences. This is not as paradoxical a claim as it may look to be at first, as the very different ways we find of framing discourses about the natural world in antiquity can shed a revealing light on otherwise invisible assumptions and problems in the modern debates, and, at the same time, the modern debates shed some much-needed light on our categorization and understanding of the ancient sciences. History informs philosophy and philosophy history.

In the case of Roman sciences in particular, though, the question finds itself in doubly sharp relief, as the very category of "Roman science" may need to be established in the first place. To many, just the invocation of the phrase *Roman science* will look like an oxymoron: the Greeks did science and the Romans did technology, or administration, or empire: Aristotle versus aqueducts.[2] What science the Romans did have was really Greek science. The Romans were, at best, "popularizers and encyclopedists," to borrow a subheading from a standard textbook.[3] Even if we try to claim for the Romans more innovation than this picture allows, even if we think some of them did some pretty good science, we are still faced with the fact that the science they were doing was largely indebted to Greek science, and there is a perfectly good word for it: Hellenistic. But is this all we can say about the science we find at Rome—that it is just a warmed-over Greek science? The answer I give in this book is: emphatically not. Not only is there more to the science that the Romans did have, but there are important aspects of what is usually thought of as Hellenistic science that are unique and distinctively Roman contributions.

To get stuck on the Greek roots of the Roman sciences is to put too much emphasis on beginnings: to ask where the Romans got their sciences from and then to run back to that source as though chasing a hare through the brush in order to find out where it lives. It lives, in fact, right where we saw it: out in the fields and forests. It may have been born in a hole underground, it may run back there when flushed, but that is not where it "lives" except perhaps in the superficial and idiomatic sense of "where it sleeps." Similarly, Greek science did not stay Greek for long. It ventured out into the great wide world and *it changed.*

By the first century BC, when the story of this book begins, and for hundreds of years after that, Greece was in a complex and difficult set of cultural and political relationships with a dominant culture very unlike itself, that of Rome. Over the previous century and a half or so, the Romans had gained increasing authority and eventually dominion over the entirety of the Greek-speaking world. But even as we say this, the words of Horace ring inevitably in our ears: captured Greece took its great conqueror captive as well, bringing in its train a host of already highly developed sciences, philosophies, and aesthetics.

In that light, what would it mean for a Roman science to be Roman? Sometimes the answer is relatively easy, but potentially thin. What, for example, do we make of Lucretius' great *De rerum natura,* written in Latin but for the express purpose of explaining a Greek philosopher to a Roman audience and, as has recently been stressed, of addressing contemporary Roman social and political ills—mostly Greek in philosophical content, perhaps,

2. Compare Moatti 1997.

3. Lindberg 1992; see also Russo 1997; French and Greenaway 1986; Stahl 1962; contrast most recently Doody 2009.

but Roman in form and context?[4] I argue in this book that form and context matter considerably, particularly when the larger frames (ethics, politics, religion) begin to indelibly shape the content as well. To take another example, consider Cicero's massively influential *Dream of Scipio.* Again in Latin, again for a Roman audience, but this time only very loosely modeled on a Greek original (the "myth of Er" at the end of Plato's *Republic*). For his part, Cicero, like Pliny the Elder and many other Latin authors, did not see themselves as merely commenting on, clarifying, or popularizing Greek originals. They saw themselves as building on them, and building something considerable at that. To quote Cicero,

> non quia philosophia Graecis et litteris et doctoribus percipi non posset, sed meum semper iudicium fuit omnia nostros aut invenisse per se sapientius quam Graecos aut accepta ab illis fecisse meliora.[5]

> [I]t is not that one could *not* learn philosophy from Greek writings and teachers, but it has always been my opinion that we Romans found out everything for ourselves more wisely than the Greeks did, or else improved the things we got from the Greeks.

A *Roman* World

Cicero's opinion notwithstanding, the reputation of Roman philosophy has certainly had its ups and downs over the centuries. The importance of Cicero and Seneca for early modern philosophy, for example, is difficult to overstate. But then their influence peaked, only to hit a low ebb beginning with the Hellenophilia of the Romantics and continuing through most of the twentieth century. Roman philosophers were thereafter often seen either as simply derivative or degenerate, or as little more than sources for the mining of the ideas of the Hellenistic greats, raw ore from which to refine the lost Chrysippus, Posidonius, or Epicurus. In the last twenty years or so, though, the reputation of Roman philosophy has been on a well-deserved upswing. We now have essay collections and monographs dedicated to the question as a whole, or to individual Roman thinkers, and philosophy at Rome is look-

4. On Lucretius and politics, see Long 1986, 2003; Fowler 1989. For a wonderful overview of the question of attention to Greece vs. attention to Rome, see Fitzgerald 2000.

5. Cicero, *Tusc.* I.1. Italics in translation mine. See also e.g., Cicero, *Rep.* III.5f.; On Pliny see Beagon 2005, 50f. For what Roman and Greek interactions meant for ideas of culture and identity during our period, see Woolf 1994; Wallace-Hadrill 2008; Madsen 2009.

ing considerably less derivative and considerably more interesting in both content and social setting than it has looked for many decades.[6]

Many of the texts we will look at in this book are written in Latin by ethnic natives of the Italian mainland. Some Latin authors are from farther afield in the empire, and others fall somewhere in between. Still others have fallen prey to the centuries, so that we know nothing of their places of origin. But even Roman writers do not always stick to Latin. Rome's very first historian, Quintus Fabius Pictor, writes in Greek, and does so as far back as the late third century BC, which shows how very early Greek cultural influences were bearing on upper-class Romans.[7] Many Roman philosophers also write in Greek, at least partly out of a feeling that Latin was inadequate for the task of philosophy. In the primary period covered by this book, Cicero, Lucretius, and Seneca all somewhat stubbornly make a point of writing in Latin, and each of them shows acute awareness of the difficulties of writing philosophy in their native tongue (indeed, the astrologer Firmicus Maternus was still openly worried about Latin's adequacy as late as the fourth century AD).[8] Cicero, for example, frequently sees the need to coin new words and sometimes has to stretch Latin syntax to make a point that may have been easier, or at least more familiar to his readers, in Greek. (The range and longevity of many of the new words coined by Cicero are frequently commented on by modern scholars, and even a partial list—of just the English words whose roots he gave us when he Latinized Greek philosophy—is indeed remarkable: moral, quality, evidence, convenience, indifference, essence, humanity.)[9]

Other authors treated in this book are not, however, ethnically Latin, nor do they even write in Latin. They are Greeks, writing in Greek—so why call their science *Roman*? One reason for doing so is in order to draw attention to the historical, social, and cultural loci in which science is happening. This approach readily acknowledges the fact that often the bearers of the science that the Romans had access to were not themselves Romans. We need to remember, though, that the Greeks who brought their sciences to the Romans were not dead Greeks, but living individuals who came physically to Rome, who corresponded with Romans, who were the beneficiaries of Roman patronage at home or abroad. Science as the Romans knew it came on the lips of and in the books brought by these foreigners. It crystal-

6. See, e.g., Griffin and Barnes 1989; Maurach 1989; Powell 1995b; Barnes 1997; Barnes and Griffin 1997; Sedley 1998; Morford 2002; Long 2003; Inwood 2005.

7. See Wallace-Hadrill 2008, chap. 1, for a discussion of what such "bilingualism" implies.

8. Firmicus Maternus, *Math.* IV, pr. 5.

9. See Powell 1995a.

lized for Romans in conversation and debate with them and with each other. And the cultural, political, rhetorical, and social contexts in which that exchange was happening were those of the dominant power, of Rome. What we might loosely call funding structures, negotiations of prestige, career advancement, networks, publication, performance—all of these happened in the Roman cultural arena. Almost none of the main Greek authors treated in this book "stayed home."[10] The career of the great physician Galen is a perfect example.[11] Born to an aristocratic Greek family around AD 130 in Roman-ruled Pergamum (Asia Minor), Galen studied medicine, philosophy, and rhetoric there before continuing his studies at Smyrna, Corinth, and Alexandria, in a kind of second-century educational "grand tour." He then took up a thoroughly Roman position back at Pergamum: he became a physician to the gladiators. But it is only with his big career move in AD 161 that Galen really jumps into the heart of things and begins his monumental climb out of provincial obscurity: Galen goes to Rome.

In Rome, Galen entered the fray of competition with his fellow physicians in earnest. One can read his delightfully self-promoting little book, the *Prognosis,* as a story of his being recommended from one important patron to another, moving higher and higher in the Roman social order with a series of very successful (and if we are to believe him, much talked-about) cures.[12] His frequent and popular anatomical performances at Rome were an important part of his claims to knowledge, authority, and prestige, and they are part of a much larger cultural trend in the second-century empire that prized virtuoso rhetorical performances—part of the cultural wave now

10. Ptolemy is the reason for the qualification "almost." We know nothing of his biography, except that he seems to have done most of his work at or near Alexandria. That he was a Roman citizen is indicated by his *praenomen,* Claudius, but we can say little more about his "Romanness" (something similar applies to Sextus Empiricus), although it is perhaps worth mentioning, as James Evans reminded me, that Ptolemy chooses as epoch for his own observations year 1 of the reign of the Roman emperor Antoninus Pius. Alexandria's place in the empire was second only to Rome's in terms of size and wealth, and the fertile mix of cultures represented in one of antiquity's most ethnically diverse cities is mirrored in the variety of sources Ptolemy draws on. He cites Caesar and Varro alongside Greeks, Egyptians, and Babylonians in his works (see, e.g., Lehoux 2007a).

11. A good recent treatment is Mattern 2008. For questions of Greek identity and Romanization, see, e.g., Woolf 1994; Wallace-Hadrill 2008; Madsen 2009. Compare also Woolf 1998; Revell 2009.

12. As has often been pointed out, Galen's constant and sometimes tedious self-praise needs to be handled with caution. His own ego aside, though, it is clear that he was astoundingly successful at what he did, so while particular details in what follows may well have been embellished by the master of self-promotion, the overall fact of his rise to prominence is not in question. On patronage and physicians at Rome, see Mattern 1999, 2008.

called the "second sophistic."[13] But the anatomies were not his only public displays of virtuosity. We see him practicing his medicine in an environment very different from the relative privacy of the modern doctor's office. We often find Galen consulting with patients in the midst of (sometimes fierce) competitors, as well as among the family and friends of the sick person.[14] He visits the deathly ill philosopher Eudemus frequently, and he is almost always, it seems, in company. The medical advice Galen gives Eudemus is not only talked about among his competitors, but is actually given in their presence.[15] When Galen heals Eudemus successfully, the effect is both public and spectacular, for Eudemus then praises Galen to all of his visitors (and visiting as a social institution is worth remarking on as an important part of the formalized Roman social networks known as *patronage* and *amicitia*). Eudemus recommends Galen to powerful acquaintances and Galen begins doing the rounds of upper-class Roman households, curing one patient after another (again, in the public eye).

Galen relates the story of his cure of Diomedes the rhetorician, who even "the most renowned" of the court physicians could not help.[16] He gives us his great feats of deduction: The case of the insomniac woman (who turned out to be secretly infatuated with a dancer); The case of the rich man's slave (who was ill from fear of the auditor); The case of the ex-consul's son (who was sneaking untimely food). How did Galen do it? ἐκ γὰρ κοινῶν ἐπιλογισμῶν εὑρίσκεται τὰ τοιαῦτα,[17] he says to his dedicatee, which one could almost translate as "Elementary, my dear Epigenes"—it would at least capture Galen's powerful rhetorical development of mystery and danger culminating in dramatic revelation and successful resolution.

Similarly, Galen offers a highly dramatized version of his cure of the ex-consul Flavius Boethus' wife, who had a complicated "female flux" that baffled all the most prestigious doctors. The upshot of this famous case, Galen tells us, is that he was recommended by Boethus even to the emperor, Marcus Aurelius, himself. Galen had, we might almost literally say, arrived

13. Kollesch 1981; Bowersock 1969, 1974; Barton 1994a; Gleason 1995; Von Staden 1997; Borg and Borg 2004. For worries about the "second sophistic" as a periodization, see Goldhill 2001a, 14. For a view of the second sophistic (and particularly of Galen's participation in it) as a fundamentally Hellenic reaction to Rome, see Swain 1996; Flemming 2007b offers an (I think good) argument against Swain.

14. See Mattern 1999.

15. See Mattern 2008, esp. chap. 3.

16. Diomedes' social standing is notable by the fact that οἱ εὐδοκιμώτατοι ἐκ τῆς αὐλῆς should be bothered with him (Galen, *Prog.* 5.3).

17. Galen, *Prog.* 6.14.

at the very heart of the empire. From this point on, Galen's narrative in the *Prognosis* never once loses sight of the imperial court, as though the rest of his future were determined by the very important persons whose names he so casually begins dropping. Eventually serving at the courts of three successive emperors, Marcus Aurelius, Commodus, and Septimius Severus, Galen was clearly very successful in the cutthroat social and philosophical competition of the imperial capital. His public displays, his speeches, his debates, his published books, none of these can be fully understood without keeping one eye on them as deliberate moves in the power game as it was played at Rome in the second century AD.[18] Galen was born a Greek, to be sure, but his success in the imperial capital shows his career to have been thoroughly shaped by the city of Rome, its intellectual culture, its politics, its rhetoric,[19] its patronage.

A Roman *World*

The main theme of this book, though, is about what it means to understand a *world*, and these social aspects are only part of that picture. If we look to the Roman sources, we find an exceedingly rich and complex tangle—every bit as rich and complex as our own, but very, very different. Sometimes startlingly so: different entities, different laws, different tools and motivations for studying the natural world. So, too, different ways of organizing knowledge, and sometimes different ways of understanding even the most basic levels of sensory experience. This book is an inquiry into how and why the Romans saw things differently than we do, or to put it more pointedly, how and why they *saw different things* when they looked at the world. Much rests on the interconnections between what may at first look like widely divergent fields of knowledge, and on debates and disagreements among the Romans themselves about what they saw and what it meant.

To take an example: in *The Dream of Scipio,* long one of antiquity's most influential prose works, Cicero uses the physical geography of the heavens as he knew it for the literary setting of a journey by two giants of the Roman historical imagination, Scipio Africanus (the long-dead iconic hero who had finally defeated Hannibal in 201 BC) and his grandson, narrator of the story, Scipio Aemilianus (destroyer of Carthage at the end of the third Punic War half a century later). Africanus leads the younger Scipio on a fantastic trip up into the sky, from which height they survey the earth below and the

18. On the relationships between performance and text at Rome, see Habinek 1998.

19. See Flemming 2007b.

circles of the planets up to the highest heaven. Not quite science fiction in the modern sense, to be sure, but science *in* fiction all the same. As it is normally read from the vantage point of the history of the sciences, we are meant to notice the descriptions of the spheres in which the planets travel, the geocentricity of the cosmos, the prescient descriptions of the stars as both larger and more numerous than our weak eyes could imagine from down here on earth. We get some geography—the classical division of the earth into climactic zones—but also a unique and humbling perspective on the size of the Roman empire when seen from a cosmic height. We are treated to an introductory lesson in the ancient science of harmonics, bundled with a description of that old premodern standby, the music of the spheres. But where the modern reader marks off the "science" bits from the rest of the text tells us a lot about his conceptualization and categorization of that thing called "science." Most importantly, we know that Cicero and his contemporaries would not have drawn the distinctions in the same ways as we would, had they been given the task.

We also need to pay attention to how the drama of the *Dream of Scipio* operates in a larger historical space, one where its author was himself pursuing political aims both within and outside of the dialogue to which the *Dream* was originally appended: Cicero's repeated emphasis on political duty to the republic is aimed directly at his own contemporaries and the profound political instability incessantly threatening the state in his lifetime. The aims and analogies of the *Dream* cannot be separated out from this external frame, and so we need to look to contemporary political, rhetorical, philosophical, and educational contexts to shed light on the multifaced ranges, responses, and interdependencies of Cicero's knowledge.

Worlds are big, both physically and conceptually. They are also highly intertwined, and often subject to debate about particulars. Unpacking Cicero's one little fictional dream, we see that it is nigh impossible to untangle the different threads that weave the narrative together: astronomy bleeds into ethics, ethics into politics, politics into theology, theology into mathematics, mathematics into harmonics, harmonics into astronomy again. Psychology plays in a minor key throughout. The more we think our way into the details, the more tangled and intersected the web in which we find ourselves. We emerge in the middle of a multidimensional world, where politics cannot be cleanly distinguished from conceptions of nature, and neither of them from conceptions of the gods. This tangle, far from impeding the investigation, instead determines our focus. The contexts, I argue, are everything.

Nature, law, the gods: a threefold cord. We cannot understand the Roman sciences except as a working out of the relationships between these three, in the face of the evidence provided by experience. But even as we say that, we need to be aware how messy a thing experience really is, and of how complex, and historically specific, its handling as evidence turns out to be.

Knowing Nature in the Roman Context

I have been using the word "science" rather frequently, and it would be worth pausing for a moment to clarify what I mean by it and what I am and am not claiming for the Roman intellectual project so labeled. I cannot pretend to offer a comprehensive definition of science, one that would fit all relevant fields, all relevant historical periods, and all relevant practices and practitioners. The word "science" in our modern sense famously arrived very late in the history of what we now call the sciences, not effectively distinguishing itself from "philosophy" or "knowledge" or "learning" until well into the nineteenth century. Nevertheless, there is a very common, though strictly speaking anachronistic, use of the word "science" to describe the subject matters of several fields of modern scholarly activity (the history of it, the philosophy of it, e.g.). It is in this vein that I have used the word "science" to categorize Roman projects aimed at understanding, questioning, and testing the natural world (I thus ask the reader to indulge my use of "science" in the phrase "Roman science" as being meant in the same spirit as Steven Shapin's use of "Scientific Revolution" when he humorously said that "there was no such thing as the Scientific Revolution, and this is a book about it").[20] To be sure, Roman projects aimed at understanding the natural world are exceedingly diverse, and I nowhere claim in this book to be uncovering one single monolith that we can call "Roman science." I have tried to capture the flavor of this point by pluralizing the word wherever possible: "the Roman sciences." There are many features of the modern sciences that do not have clear ancient analogues. Fields are differently defined, and the boundaries of the natural world itself (both conceptual and physical) are not even static over time. Borders and overlaps between science and technology, or between these and literature, need to be very differently handled by the historian.[21] There were no ancient universities, no scientific conferences,

20. Shapin 1996, 1.

21. There has been considerable recent work in French, Italian, and German on what is called ancient "technical literature," a designation meant to include works on agriculture, artillery, strategy, surveying, aqueducts, medicine, and more (see, e.g., Amouretti and Comet 1995; Nicolet 1996; Amouretti 1998; Meissner 1999; Pigeaud and Pigeaud 2000; Formisano 2001,

no journals where investigators published their results. So, too, no *New Scientist*, no science pages in the *New York Times* where the newest work could be reported, compared, commented on. From these modern sources there often emerges an understanding, among professionals and among the scientifically literate public, of something we might call the "consensus in the field" on many issues. Do humans cause global warming? There seems to be a consensus that we do. Do prions cause bovine spongiform encephalopathy? It looks like it. In antiquity, though, the practitioners, the writers, the thinkers about nature were more diverse, less collectively organized, and considerably less in agreement on many issues. They also did not use consensus in order to inform policy in the same way that we do. This point matters if I am to be clear on what I mean by "worldmaking" in the subtitle of this book. I do not mean to say that the Ptolemies and Galens, the Ciceros and Senecas of my study are passing down to the Roman in the street what the truths about nature are, that they are the arbiters for an entire culture on all matters of nature. Instead the sources in this book are every one of them participants in philosophical debates about nature, debates that are happening at a culturally elite level. This situates what they believe—and what they know—about nature at one remove from the vast majority of the people working the fields and baking the bread that kept the empire thriving. To be sure, those people, those women, slaves, farmers, merchants, all of them also surely had much to say on how nature works, why one kind of manure is good for grapes and another good for vetch, why wine ferments better at one temperature than another, why a mullet caught under a rock tastes better than one caught in open water.[22] They are sometimes them-

2003, 2004; Long 2001; Santini et al. 2002; Horster and Reitz 2003; Peachkin 2004; Paniagua Aguilar 2006). Much of this work has tried to situate *Fachliteratur* within its contemporary social, intellectual, and literary traditions (perhaps better: "literary space," on which see Cavallo et al. 1989-). This approach has two important effects that bear at least some comparison to the present work: the first is the importance of cultural situation. Roman technical literature is not just bad Greek mechanics or architecture, nor is it a poor attempt at literature generally. At the same time, though, the *Fachliteratur* scholarship (as indeed the literature that it studies) often tends to center on work that does not bear directly on the epistemological concerns of the present project. Science and technology tend to be seen as running hand in hand in the modern world, and I think a good case can be made for a more important relationship between theoretical and practical concerns in antiquity than is commonly acknowledged, but that would be a question for another book (see, e.g., Berryman 2009; Tybjerg 2005; Schiavone 1996).

22. Not that this class of people could have afforded to eat anything so outrageously expensive as a mullet, but they were the ones who caught, prepared, and served the fish to those who could.

selves also the objects of scientific scrutiny and normativity. But this is not their story, interesting and important though their stories be.[23]

Instead, I mean something special by worldmaking, something I have borrowed from philosophical debates on the subject of whether and to what extent the world is constructed by, or contingent on, a society's ways of approaching, organizing, compartmentalizing, and understanding it. The authors I look at in this book are members of a vibrant, if sometimes volatile, class and culture, who have much in common over time, even if we can never lose sight of just how very much changes over the several centuries I treat in this book. Here again, I try very carefully not to generalize about *a* Roman science, just as one would be wary of generalizing about an American science over an equivalent length of time (extending from 1690 to the present, say). Instead, the philosophical themes I explore pick out certain authors, certain texts, because they best illustrate the themes and arguments that I am making as I build the larger argument of the book. The frequent recurrence of Stoic and Platonic themes reflects their broad predominance in the period, to be sure, but I should emphasize that, in this book, I do not sketch a historical trajectory, a synthetic developmental arc of the Roman sciences. Instead, I use episodes from Roman scientific authors to build a historiographic and philosophical case about the integration of observations and worlds, ancient and modern. Chapter by chapter, and sometimes across two or three chapters, there is much history to be found, but the book as a whole is not a synthetic history per se.

So what is this thing called Roman science? What are its methods and objects of study? How is the world parceled up into comprehensible packages? How is knowledge about the natural world obtained? How is it passed on? How is it refined? As soon as we begin to dig into the details of their sciences, several things emerge. No sooner has our spade broken ground than we notice that the Romans are dealing with some entities, either as objects of study or as explanations, that do not even come under the scope of what we now call science. The biggest, the most sweeping of these, are the gods. An interesting tension then emerges, insofar as modern scholarship has so often touted the exeunt of the gods-as-causal-actors as one of the hallmarks of what it means for an ancient explanation to be *scientific*. Recent scholarship has begun to argue that this is a mistake, and my analysis takes

23. For work that looks explicitly at the ancient sciences and nonelites, see, e.g., King 1998; Flemming 2000, 2007a; Cuomo 2001, 2007.

very seriously some of the theological concerns that play into—and emerge from—investigations into nature in Roman antiquity.[24]

The other set of concerns central to an understanding of Roman approaches to nature comes out of their legal contexts. Again and again in this study, we find that concepts of law are at the very heart of how the Romans approach nature. But just as it has been commonly supposed that the gods exited the history of sciences on day one, so it has also been supposed that lawlike explanations of nature only entered the history of science at a particular point, usually situated by modern scholars sometime around AD 1600. By carefully delimiting what it might mean for an explanation to be lawlike, and by looking at exactly how law and nature come up against each other in ancient sources (and this necessarily includes both Roman and non-Roman sources), we find that far from maintaining some essentialized distance, law and nature should be seen as saturating each other in the period we will be looking at. Indeed, as we explore the nature of this saturation and its multiple implications for the epistemology of experience, we find not only lawlike descriptions of nature, but also extensive uses of nature in normative, ethical, and political frameworks. Nature has ethical implications, but also—again surprisingly—ethics has natural implications. Observation, that most basic of tools for knowing the natural world, is shown to be ethically loaded.

In exploring the ways in which the observation reports of others are framed in this historical discourse, we see that by the time of the early principate, much of the epistemological work being done in the Roman sciences is not happening under the exclusive eye of philosophy or of logic, but is happening under the auspices of a different discipline with a different set of tools for knowing, the discipline the Romans called *rhetoric*. By the imperial period, rhetoric had long been the undisputed backbone of any upper-class education, and the tools it had developed for adjudicating difficult cases, contradictory witness claims, and probable—but not logically necessary—arguments, these tools are used to powerful effect in framing arguments about the natural world. The model of judgment, literally understood, comes to form a cornerstone of the epistemology of sensory experience. To be sure, jurisprudence had left its stamp on even the earliest Greek philosophers and physicians,[25] but we see in the Roman context a much more comfortable relationship with, and deliberate use of, the rhetorical

24. See, e.g., Beagon 1992; Horstmanshoff 2004; Flemming 2003; Sedley 2007; Taub 2008.
25. See, e.g., Lloyd 1979; Detienne 1981; Asper 2007.

tool kit.[26] In particular (and I think quite remarkably) we see by the imperial period a self-conscious use of forensic rhetoric—modeled on contemporary courtroom practice—in the examination of the reliability of sensory evidence and of eyewitness testimony. Indeed, I suggest in chapter 4 that we can read Seneca's remarkable *Natural Questions* as belonging to a then-popular genre of rhetorical treatise (*controversiae*) that examined difficult moral problems as though they were court cases, with the reader playing the role of judge.

As this epistemology gets fleshed out, though, we begin to notice that the gods are not the only objects of ancient knowledge that appear to the modern eye as nonexistent, "unscientific," or implausible. How *do* we take seriously all those silly monsters, the dog-headed people, the spontaneous generation, the three-foot-long grasshoppers, the astrology, the divination?[27] Here there are two questions: what are they doing there, and where did they go? I offer a reading of what we might call the life histories of these entities, which poses some difficult challenges to the still very common claims for a radical historical disjunction of scientific method, usually situated around the Scientific Revolution. Keeping a careful eye on observation and experience, experiment, testimony, and rhetoric, we will explore some of the most difficult cases—sympathy, astrology, divination, and numerology to name a few. Strange though many of these cases may seem to the modern eye, their difficulty lies not in their weirdness but in the ancient claims as to how good the evidence for them is.

There is a second benefit to this attention to the apparently impossible. By paying close attention to observation claims that we think cannot be true, we block one of the easiest and most deceptive moves in the historiography of epistemology: the assumption that people believed things—nay, even *saw* things—because they were true, because they were really out there in the world to be seen. Removing this recourse, we magnify all the other features of observations and observation claims: how they move from one person to another, what kinds of authority they have, where they stand in arguments, and most importantly for this book, how they hang together to make up a world. Long gone are the days when superstition or ignorance would work as explanatory categories here, and yet, I argue, we still do not have an adequate account of these oddest excrescences of ancient science. It would be easy enough to situate these strange entities in modern arguments about textual authority, psychology, or the sheer voluminousness

26. See Long 2003; Lloyd 1979; see also Stroup 2007.

27. Compare Li Causi 2003, 2008.

of "the unknown" in antiquity, but then the danger lies in overstressing the differences in "mentalities" or epistemological criteria between ancient and modern.[28] Instead, I explore how the world that the Romans saw themselves as occupying hung together as a coherent whole, emphasizing the very powerful and productive observational rationality at the heart of Roman approaches to nature. It is not that the Romans knew only a little and were puzzled about a whole lot, but that they thought—just as we do—that they had a pretty good idea of what was going on in the world. In saying that, though, I too face a great danger, and I do not wish to be seen to be arguing anywhere in this book that ancient science or epistemology is just the same as modern science or epistemology—a conclusion that some readers may be tempted to draw from the way I move back and forth between ancient and modern questions. It has long been a core part of literary studies of the ancient world to look unabashedly at ancient literature with the aid of modern theory. This is not to impose a foreign structure that was absent in the first place, but instead to see how the modern categories stand up to and are challenged by a body of foreign texts, and at the same time to see what new methods of reading those texts can reveal about antiquity. Studies on sexuality and mythology have benefited from a similar approach. These approaches are two-way streets, where modern theory enriches our readings of ancient sources, and where ancient sources challenge and stimulate the foundations and assumptions of modern theories. So also in the present investigation, such an approach highlights the ways in which our understanding of ancient science is revealing of our own often unquestioned criteria for understanding history or the natural world.

Are there differences between ancient and modern science? Of course there are. Are those differences fundamental? Did things change suddenly? Can we pinpoint some radically new way of doing things that emerged at some discrete point in history when we got something we call modern science? I think not. Instead, I want to see the history of intellectual engagements with the natural world as nodes in a shifting conversation. The sets of debates that we find in the history of the sciences are conversations that happen in and across time. We, now, are talking to and responding to each other and to our more or less immediate predecessors, who in turn were talking to and responding to each other and their immediate predecessors. This conversation, with shifting participants and shifting interests, can be traced back, in one important strand at least, to the debates we see taking place in this book. (Here it may be worth keeping in mind that we are only sixty

28. Against mentalities, see Lloyd 1990; Tambiah 1990.

generations or so removed from Galen.) This argues for a continuist model rather than a disruptivist one. Stark incommensurability has no part in it, nor do essentialist claims about the fundamental characteristics of modern versus ancient science, or of modern versus ancient mentalities. Not to say that it's now the same as it ever was, but the differences are deceptively easy to overstate, and the temptation to find points of rupture (whether positivist or Foucaultian) can only distort both the before and after pictures.

So what *did* the Romans know about their world? Quite a lot, as it turns out. By keeping both ancient and modern epistemological considerations at the fore, we will see that even some of the most intractable-looking aspects of their science, the ones most amenable to being flagged as fundamentally foreign—the kinds of things that have been fodder for the periodization of science into ancient, medieval, early modern, modern—even these aspects are not so incongruous or isolated as they first appear. But this historiography of continuity still has one major theoretical hurdle to overcome if it is to fly: how can we hold a continuity thesis while recognizing the very profound differences between their world and ours?

Indeed, is all my talk of "worlds" just an appeal to relativism? Questions around relativism and realism are so central that most readers will already have a position sketched out one way or the other. Some readers will be (justifiably) weary of the seemingly endless philosophical arguments about realism and relativism that fill the journals, but, I argue, the problems are still central not just to philosophy but to history as well, and the historical material covered in this book should enrich and challenge the philosophical debates.

Historians and many philosophers of science have reacted in a variety of ways to the problems Thomas Kuhn famously raised, which has often led to what appears to be either an explicit or implicit relativism about scientific entities, or else to what amounts to a phenomenological *epoché,* a bracketing of the question about what is *really* out-there. I argue that both relativism and phenomenological bracketing present serious historiographic problems, insofar as an important part of how we understand ancient science in the light of Kuhn's critique involves us reapproaching and rethinking our own experiences of nature in order to see nature through Roman eyes. But there is an object to those verbs (reapproaching and rethinking) that must be more than just grammatical convenience for the method to work at all.

Rocks and planets, birds and bees figure large in this story, but their roles, like Falstaff's, are not limited to only one play. The ancients talk about rocks, and I think I can understand them. But I think I can understand them not just because I have read their writings on rocks carefully, but also because I think I know something about rocks from my own experience, and that experience both helps and hinders my attempts to understand their ideas about the world. This is the great methodological tension that inheres in any investigation of historical science. Heraclitus said that the sun is new each day. In some senses I would agree. But the point here is that there is still a sun, and we are still people trying to understand it. Now the question is, what do we make of the different understandings over time? Various kinds and degrees of realism, semi-realism, and antirealism are available off the shelf, but none of them are up to the historical challenge without serious and careful modification, and it is only in the light of the obstacles thrown up by our analysis that we can see precisely where and why the circle needs closing. The final chapters of the book thus use the historical material to directly engage the philosophical question about whether and how historicized scientific objects can be seen as *real,* and what that means for both philosophy and history.

By exploring a wide range of sources for what is unquestionably the most prolific period of ancient science, that from the late Roman republic to the early principate, from highly technical works by Galen and Ptolemy to the more philosophically oriented physics and cosmologies of Cicero, Lucretius, Plutarch, and Seneca, we see how ideas of nature, law, and divinity weave a complex web, and how they interact with observation and classification to produce something called *knowledge,* about something called *the world.*

Overview

I think of the argument of this book not as a linear trajectory, but as an interconnected set of concerns that map out a territory of exploration. This means that the ordering of the chapters could have been framed quite differently, but in the end it found its current formulation because this ordering allows for at least one fairly straightforward path through the interconnected questions asked. Accordingly, the chapters are arranged following broader philosophical themes instead of a linear chronology. I try as much as possible to flag chronological shifts as I move through the argument, and to be attentive to the profound historical changes that took place over the three centuries covered here, but I do still ask a little patience of the historian to allow me to move forward in time and then back again as the larger argument demands it.

The book does, however, begin more or less at the beginning, at a time when we first see Roman intellectuals engaging with Greek philosophical texts in a sophisticated and sustained way, which is to say during the first century BC.[29] Choosing Cicero as my starting point, I argue that the engagement with nature in his thinking is very closely related to his concerns in politics and law. One important aspect of this is the way he uses nature to ground the political system for which he is desperately fighting in the last days of the republic (nature as politics). That he is thinking of nature in this legal and political context opens several prominent questions, not least of which is how nature and law could have come to intersect in the first place—after all, did Cicero not inherit the old Greek separation between *nomos* (law) and *physis* (nature)? Chapter 3 looks more closely at the question of the relationships between law and nature in antiquity, arguing that in fact a strong distinction between the two is far from universal, and that law and nature come together very comfortably and very deeply in several different ways in ancient sources, Greek and Roman alike. Part of this relationship gives the lie to one common marker of periodization in the history of the sciences, one that says that "laws of nature" are a relatively modern concept, invented in or about AD 1600.

Building on this, in chapter 4 I will try to show not just how jurisprudential ideas interacted with thinking on nature, but how actual courtroom practice came to the aid of epistemology by creating a space for testimony and eyewitnesses in assessments of probability (here used in the very general sense of "what is probable," not in the modern technical sense of mathematical probability). But the reliability of eyewitness claims is evaluated, again following standard rhetorical practice, in large part according to the ethical standing of the witnesses themselves, and so moral particulars come to underscore and guarantee the legitimacy of observations of nature. I focus in this chapter primarily on Seneca's (first-century AD) often underrated and still understudied *Natural Questions.* In the process of this exploration of the reliability and probability of witnesses and testimony, light begins to be shed on difficult problems inherent in the epistemology of observation generally, problems that will form the core of the next two chapters. Chapter 5 thus goes on to look at one set of strategies employed, in the second century AD, for trying to lock down observational certainty (then under serious threat from Sceptic attacks) through the careful exploration of the actual physical and physiological pathways of observation. I sketch

29. We know that some Romans took an interest in Greek philosophy at least a century or so earlier (the Stoic philosopher Panaetius was part of the circle of the second century BC Roman general and statesman Scipio Aemilianus, for example), but the first century marks the first time we see Romans writing philosophy for themselves.

two concerted forays into the problem of certainty in observation with twin prongs in mathematical optics and in anatomy. We also notice, though, that important points are offered in the ancient accounts where we, as outsiders looking back at a very foreign set of ideas, see more clearly than the participants in the ancient debates themselves did, that nontrivial assumptions are being made about mechanisms or laws of causation. The point that emerges, though, is not limited to what ancient observers did or did not see, but turns out to be a problem with how observers encounter the world more generally. The "black boxes" that get highlighted here are our first window onto the epistemological blind spots that dog observers, ancient and modern alike. Chapter 6 continues the exploration of these blind spots, arguing that they are ubiquitous and inevitable concomitants of how observers approach the world. We also see that relatively innocuous-looking assumptions about how phenomena are related, and how those relationships enable possibilities for interaction, can have major effects on how the world itself looks to be put together, and on what kinds of things are possible or impossible, patently obvious or patently ridiculous, in that world. Chapters 7 and 8 continue to explore these rippling effects, looking first at how the reality of certain kinds of entities can scale up to have major implications for the world at large, and then how assumptions about the order and intelligibility of the cosmos scale down to affect everyday objects as they are conceived and seen in observational practice.

Having thus taken multiple fronts on which to problematize the observational world, I end the book by stepping back and asking the larger philosophical question of whether the questions and challenges raised by our look at Roman ideas about their world should drive us to a philosophical relativism or scepticism. In chapters 9 and 10, I set out reasons why a certain degree of realism seems inescapable for the historian, most prominently because of how the historian's own experiences of the natural world figure into how we make sense of our historical subjects and their experiences of that same world. I begin, though, with the argument that realist claims in historical sources are much more difficult to cordon off than philosophers generally recognize. The threat of the history of science to some key realist arguments has long been recognized (philosophers of science will be familiar with Larry Laudan's refutation of Hilary Putnam's "miracle argument").[30] I argue that standard attempts by philosophers to cordon off the problems posed by such historical counterexamples fail, and they fail inescapably. The solution lies in finally allowing our historical actors their

30. Laudan 1981a; Putnam 1975, 1978.

realisms, although I acknowledge that this puts significant pressure on what we can mean by *real.* In chapter 10 I show how that pressure can be relieved by looking very closely at the conjunction between *real* and *true.* By turning from naïve correspondence theories of truth and reality, and looking instead to epistemological coherence and pragmatist truth instead, I argue that we can sketch an epistemology of nature that is consistent not only with historical scientific change, but also with the historian of science's often unacknowledged use of experiences of the world as an integral part of the historiographic tool kit.

CHAPTER TWO

Nature, Gods, and Governance

Roman ideas, assumptions, and intuitions about nature, and indeed their very reasons for studying and theorizing about nature, stand in an intellectual context that is unique to its time and place. This is not to say that the Roman context was static over time—it was not—but to suggest that part of the answer to the basic question of what makes the Roman sciences *Roman* is necessarily rooted in their situation relative to Roman culture as a whole. As we will begin to see in this chapter, Roman ideas about nature interact in unique ways with Roman ideas about theology and politics. For most Romans, as for most ancients, their understandings of the natural world came framed by, and saturated with, theology. After all, a belief in the gods was very nearly universal among ancient authors, even those (the atomists) whose purely mechanistic universe did not require divine planning or intervention. So, too, there seems to have been a widespread belief that nature showed the beneficence and magnificence of the gods. We explore in this chapter how a theology rooted in the beneficence of the gods used a sophisticated understanding of nature to try to legitimate the Roman republican system of government at a crucial (and in the end, fatal) moment in the republic's history. The ways in which many, if not most, upper-class Romans understood the gods to act in the world, and coupled with this the ways in which the Roman state saw its own legitimation *vis à vis* the gods, together determine a larger theoretical framework within which the Roman sciences situated themselves, a framework that has *ethics* as one of its central structural supports.

The main textual focus of this chapter will be a group of dialogues by Cicero, most centrally his great theologico-philosophical trilogy *On the Nature of the Gods, On Divination,* and *On Fate* (the last extant only in fragments). Written in the late 40s BC, they are among the earliest works of

Latin philosophy. Once profoundly influential—Voltaire called the first of these works "perhaps the best book of all antiquity"[1]—they have only recently begun to reemerge from relative obscurity as significant philosophical works in their own right. By looking at them as sustained arguments and paying close attention to their historical and intellectual contexts, we will begin to explore in this chapter an important set of resonances between nature, the gods, and ethics that come together in unique ways at this particular juncture in history. I will argue that divination—the practice of expecting messages from the gods or answers to direct questions in the form of signs in the world around us—along with its theological and political commitments, offers us an important window into Roman understandings of nature and the natural world. So, too, we will begin to open up a broader set of questions for the book as a whole surrounding the uses of observation in arguments about nature.

Divinity and Divination

Although cultural historians and historians of religion have had a good deal to say on Roman divination,[2] historians of the sciences have only in the last half-century or so begun integrating divination into their discourse as anything other than the kind of magical thinking that science struggled in its initial stages to define itself against. In the history of the Mesopotamian sciences, attention to cultural and educational contexts, as well as literary form and content, has meant that since the 1970s the histories of Mesopotamian science and divination have become inextricably intertwined.[3] Work by Geoffrey Lloyd has also been productively rethinking relationships between divination and science in Greek culture.[4] In the Roman context, however, the relationship of divination with the history of Roman science remains to a large extent still obscure. Occasionally and very helpfully for our purposes, philosophers have looked in some detail at arguments on divination in (especially) Cicero, Sextus, and Plutarch, and the sophistication of the various positions the ancients hold is often impressive.[5]

1. ". . . le meilleur livre peut-être de toute l'antiquité," Voltaire 1771, 112. In second place, Voltaire put another late work of Cicero's, *De officiis*.

2. See, e.g., Johnston and Struck 2005

3. See, e.g., Reiner and Pingree 1975–98; Oppenheim 1977; Reiner 1995; Hunger and Pingree 1999; Swerdlow 1999; Koch-Westenholz 1995, 2000; Lehoux 2003b; Rochberg 2004; Ritter 2005.

4. Lloyd 1979, 1990, 2002.

5. See, e.g., Long 1982; Denyer 1985; Schofield 1986; Cambiano 1999.

Cicero's *De divinatione* is *the* great philosophical critique of divination in antiquity. It has often been taken for granted that the *De div.* stands as a sceptical attack on the superstitious practices of divination. We are told by various modern commentators that Cicero thinks that divination is a "convenient fiction"—"merely fraud and politics"—as "unreasonable" as it is "deceitful."[6] *De divinatione* stands as a "vigorous rationalistic protest" against a "superstition" upon which Cicero heaps "ridicule."[7] Even though I argue strongly against such a reading, I do still confess that it makes a certain amount of sense—at least on the face of it. In recent years, though, nagging doubts about this naïve interpretation have started to accumulate. This is particularly relevant in light of work in cultural and political history and the history of philosophy since the 1980s. It was at that time that the first serious challenges to the prima facie position were volleyed by a group of Cambridge scholars (Mary Beard, Nick Denyer, and Malcolm Schofield), whose work stands as a watershed for recent rethinkings of the cultural, religious, and political roles of Roman divination.[8] The puzzles, though, still run deep.

The *De div.* begins with a spirited defense of divination in the mouth of Cicero's brother Quintus. In the prologue to the actual argument of the book, Marcus (Cicero himself, inserted into the dialogue as a character)[9] begins by telling us that everyone everywhere has always believed in divination, and that it has been an important part of Roman culture and politics since the beginnings of Roman history. The majority of philosophers have approved of it (if to varying extents), and there are many arguments advanced in its favor. Nevertheless, Marcus thinks the time has come for a careful rethinking of the question, in no small part because divination played such a central role in the Stoic arguments in the *De natura deorum*, the dialogue immediately preceding the *De divinatione*. The basic form of the most common Stoic argument for divination is given by Cicero as follows:

6. Pease 1979, 10–11; Linderski 1986, 335; Falconer 1923, 220; Guillaumont 1984, 9.

7. Pease 1979, 12–13; Falconer 1923, 220.

8. Linderski 1982; Denyer 1985; Beard 1986; Schofield 1986; North 1990; Potter 1994; Johnston and Struck 2005; Kany-Turpin 2004; Fox 2007.

9. Beard 1986 warns us against attributing *every* opinion voiced by the character Marcus directly to Cicero himself as though it were "what Cicero really believed on divination." Just as we will see with Seneca's use of the *controversia,* the rhetorical form of the dialogue has more to do with putting arguments before the reader than it does with representing true-to-life versions of its character's opinions. To play it safe, I will follow Beard in referring to the author of the dialogue as "Cicero" and to the character in the dialogue as "Marcus."

If there are gods and
(a) they love us, and
(b) they are not ignorant of the future, and
(c) they know that the knowledge of the future would be useful for us, and
(d) it is not beneath their majesty to communicate with us, and
(e) they know how to communicate with us,
then divination exists.[10]

The Stoics see the premise and the conclusion as bidirectional: Not just "if there are gods, then there is divination," but also "if there is divination, there are gods."[11] Quintus wants in the *De div.* to carve out an argument that is at one and the same time both more streamlined, and much messier, than the fullest Stoic package.

Ontologically, Quintus' argument is streamlined because it does not at its outset require any belief in the gods or any assent to school-based doctrines. The gods and Stoic physics may come in at the end as an explanation for what we have been convinced is the case by Quintus' argument, but unlike the full Stoic argument outlined above, we do not need to begin with the gods themselves. Quintus is effectively trying something like the inverted Stoic argument ("if there is divination there are gods") even if, so far as the dialogue goes, he leaves the specific conclusion partly understated.[12] Basically, Quintus' argument chooses not to begin from widely agreed causal premises, but instead tries an argument from enumeration—what we might anachronistically call an inductive argument—that lists instance after instance of successful prediction in order to convince us that divination is true. Although many of these examples come from Greek (and a few other) sources, the majority are carefully chosen from Roman traditions.[13]

We see a parade of examples from key points in Roman history, from the very founding of the city to the present. The authority of the group of priests known as the "augural college" gets linked to both the history and geog-

10. After *De divinatione* I.82–4. Compare also *Laws*, I.16.

11. *De div.* I.9–10. Cicero calls these two arguments the "Stoics' fortress," *arx Stoicorum*.

12. But see *De div.* I.10. For a Stoic argument from outcomes of divination, see Diogenes Laertius VII.149.

13. On historical example and anecdote in Cicero, see Fox 2007. On the use of foreign examples in the *De div.*, see Krostenko 2000, 361f. Quintus' use of foreigners is complex, sometimes simply quoting famous stories from around the world to make a point about universality, and sometimes to point to how serious the recent degeneration in the use of auspices at Rome is, when even foreigners are doing better.

raphy of the city: the staff used by Romulus to mark out the ground at the founding of Rome was still used by augurs as the symbol of their authority in Cicero's day. Augury is not just a quaint tradition to be sentimentally or cynically maintained; it is rooted in the very soil and space that the Romans occupy. The great early wars that saw the initial expansion of Roman power each get brought up by name and the famous auguries and omens associated with them paraded: The Samnite wars, the Latin war, the war with Veii. The greatest challenges the empire had ever faced also figure, with several omens from both sides of the Punic Wars, including a divinatory admonition to Hannibal not to commit sacrilege by taking a golden statue from a temple of Juno. The famous divinations associated with more recent pivotal events are also highlighted: the death of Crassus, the battles at Dyrrachium and Pharsalus, Caesar's assassination. Cicero even puts in a plug for his own importance with a lengthy quotation about the significance of the portents that presaged the Catilinian conspiracy during Cicero's (oft-self-trumpeted) consulship in 63 BC. Every stage of Roman history is represented, and key events in the determination of Roman self-image are flagged. If divination is false, says Quintus, then history itself is false.[14]

Roman Virtues

Quintus' approach is doubly messy. Many theories of knowledge in antiquity distrusted purely enumerative arguments, so Quintus leaves himself exposed to some very obvious lines of attack, which Marcus is more than happy to exploit in his response in book 2. The second messy aspect is the actual contents of his list of successful divinations. If he calls up every old story he can think of he will simply be throwing himself onto the one horn of the dilemma (here meant in its technical rhetorical sense)[15] initially posed by Marcus:

> nam cum omnibus in rebus temeritas in assentiendo errorque turpis est tum in eo loco maxime, in quo iudicandum est quantum auspiciis rebusque divinis religionique tribuamus; est enim periculum ne aut neglectis iis impia fraude aut susceptis anili superstitione obligemur.
>
> In any matter, recklessness and error in granting our assent is shameful, but most of all in this one, where the question is how much weight to

14. E.g., *De div.* I.33.
15. See Krostenko 2000.

> give to *auspicia*, sacred rites, and *religio*. For there is a danger that we will be guilty either of an *impius* crime if we neglect these things, or of the gullibility of old women if we believe them.[16]

Quintus, in short, could come off looking very foolish if he backs every silly story. But we need to be careful in saying this. For Marcus' "old women" are Roman "old women" and the measure of what counts as gullible needs to be handled carefully. Not every story Quintus will tell would be seen by Cicero's intended audience as ridiculous, whatever the modern reader may think of them, and so to know whether he is in fact being portrayed as a dupe, we need to attend carefully to how a Roman would have heard the stories Quintus recounts of prediction by lightning, dreams, livers, and the eating habits of sacred chickens. There is a little more to this admonition than may appear at first sight. What is at issue is not just our conception of how the Romans would have thought, but a very deliberate argument on Cicero's part about what should count as *Romanness*.

The clue in this immediate passage lies in the three words I have left in the Latin in my translation: *auspicia*, *religio*, and *impius*. An easy translation would have been to render these with their English cognates. Then the question would be whether to trust "auspices, sacred rites, and religion," and the danger would be "impiety." But the Latin is much more loaded than the modern English terms can convey. *Religio* is not really "religion" in the English sense, where there is so much emphasis on belief, and where one even has options to choose between different religions and sects. Instead, *religio* is something a Roman *does*.[17] It is an importantly traditional obligation for the performance of sacred duty to the Roman gods that is one half of the set of mutual responsibilities for care that the gods and humans have for each other. This has both philosophical and political dimensions. Philosophically, it raises the sticky question of religious belief versus just going through the motions, and politically it means that the functioning and structure of the Roman state is inextricably suffused with ritual performance and duty.[18] In a different context, Cicero offers us a definition of *religio* as parallel to the obligation that we owe to our parents: *in communione autem quae posita pars [sc. virtutis] est, iustitia dicitur, eaque erga*

16. *De div.* I.7.

17. By which I do not mean that there is no belief tied up in *religio*, but that duty is its primary focus. For a masterful recent exploration, see Ando 2008. Roman religion is an exceptionally complex and fruitful field of recent study. See, e.g., Beard, North, and Price 1998; Rüpke 2001, 2007; Rives 2007.

18. See, e.g., Cotta's comments at the beginning of book III of the *De natura deorum*.

deos religio, erga parentes pietas, "That part [of virtue] that is displayed in public is called justice, whereas that for the gods is *religio,* and for parents *pietas.*"[19] That these are not mere dispositions of mind or belief is clear in the next sentence: *atque hae quidem virtutes cernuntur in agendo,* "Those virtues are discerned *in action.*" This is not to say that "belief" as a category is absent in all of this, but rather to identify action as what really counts. Likewise, when the sceptical Cotta in the *De natura deorum* offers his defense of the Roman pontifical college, he carefully ships "belief" (*opinio*) as a category over to "belief in the worship (*cultus,* i.e., an act) of the gods."[20]

We see this even more clearly in the laws against sacrilege that Cicero enacts.[21] The crime that a *sacrilegus* is guilty of is emphatically not a crime of belief, but one of action. It is the stealing of anything that is *sacer,* meaning anything that has been consecrated (again: action)—exactly the crime we saw Hannibal nearly committing, had he gone ahead with his plan to steal the golden statue of Juno. Belief was in many ways parasitic on the fundamental emphasis on action and duty. As Clifford Ando has recently argued, knowledge of the gods, of their existence, of our relationship to them, of their powers and abilities, all of this was built on the observation of the efficacy of ritual.[22] Devotion is most importantly a performative virtue, not a state of mind. Philosophers, even Cicero himself, can argue about the existence or nature of the gods all they like, but in the civic sphere, action is what counts. Indeed, Cicero's contemporary Varro, the "most learned of Romans," divided *theologia* into three types: mythological (the fictive and inconsistent province of poets), physical (the province of the philosophers), and civic. Theological speculations and arguments about the nature of the gods are all well and good for Varro, but they have a place, and that place is explicitly said to be behind walls (*intra parietes*) in the philosophical schools rather than out in public (*in foro*).[23]

Similarly, *auspicium* is not just another word for "divination" in our general sense.[24] In the political sphere, "the auspices" in Latin refers to a specific aspect of state and military decision making where divinatory rites (sacrifices, sacred chickens) were performed or divinatory intercession anticipated before certain kinds of decisions could be made or actions un-

19. *De part. or.* 78.
20. *De nat. deor.* III.3f.
21. *Leg.* I.22, I.41.
22. Ando 2008.
23. Augustine, *Civ.* VI.v.
24. Although there was a common use of the word to mean *sign* in general. Nevertheless, the significant political usage pushes the word away from any clean English equivalent.

dertaken. The idea was effectively to allow the gods a say at the most important points in the political and military decision-making process.[25] The gods could be asked to say yes or no or could simply interrupt by means of a remarkable event (lightning, sudden deaths). Thus *auspicia* are not generally about predicting the future, but about securing approval from the gods before an important undertaking. Augural and auspicial rites were performed by magistrates, the holders of political office, sometimes specially appointed for the purpose (as the augurs were) and sometimes as part of a larger set of responsibilities (as one of the duties of a praetor or a censor, for example). Failure to take the auspices before important actions, such as sittings of the senate, votes for magistrates, decisions to go to war, would mean that the vote or other action had no legal footing and was therefore invalid. Cicero preserves a story about Tiberius Gracchus (senior), which he repeats both in the *De div.* and in the *De natura deorum.*[26] During the consular elections for the year 162 BC, one of the ballot collectors fell over dead in the middle of everything. The senate, sensing a divine naysayer, called the Etruscan haruspices, who reported a ritual fault. Gracchus was incensed, for, as consul and augur himself, he had been scrupulous about the relevant rituals and had made sure that the auspices had been properly observed. With a racial slur against the Etruscans, he allowed the voting to be completed, and when his term ended he stepped down as consul and went off to govern Sardinia. Once there, though, he recanted. For on further reflection he realized that in the run-up to the election he had accidentally crossed into the *pomerium,* the sacred boundary of the city, and had forgotten to take the auspices when he crossed back out again. This ritual fault meant that the consuls for the year after him had not been elected validly. On hearing this, the senate duly asked the consuls to resign. Gracchus, his successors, and the senate all get praised by the Stoic character Balbus in the *De natura deorum,* as by Quintus in the *De div.*, for their *pietas.* Moreover, the realization that the Etruscans were right about the ritual fault is highlighted as a successful instance of divination.

Like many aspects of *religio,* the auspices are not in the first instance something to be believed in, but something to be performed as an important part of the running of the state.[27] Because the auspices underpin the legitimacy of the entire Roman legal, political, and military systems, these are not practices to be rejected lightly. As for the office of the augurs, it gets

25. Linderski 1982.

26. *De nat. deor.* II.10f.; *De div.*, I.33.

27. This is to say that practice matters, not just theory. See Hacking 1983; Galison 1987.

called by Cicero the "highest and most preeminent office in the state,"[28] and a modern commentator points out that the office was "guarded by the nobility even more jealously than the consulate."[29] Cicero himself had been elected to this prestigious office, and in fact he was one of only two "new men" to have been so elected since at least the early third century.[30] Quintus plays repeatedly on Cicero's election to the augural college by using second-person plurals and possessives when talking of augurs: "you people think . . . ," "your colleague says . . ."[31]

Finally, the worry about the neglect of these important matters being *impius*. The word is simply the opposite of *pius*, which is not "pious" in the English senses of pure-of-belief or pure-of-life, but rather "dutiful." This can and does include duty to the gods, and so has a strong religious flavor, but it also includes duty to parents, patrons, and others (it thus encompasses shows of gratitude, affection, and kindness). So, too, *pietas* includes duty to the state (thus covering "patriotic duty"). The death of Scipio Aemilianus, for example, which Cicero wants to blame on the Gracchi (Scipio's cousins), is said by Cicero to have been brought about with *impius* hands, in violation of familial and political obligations both.[32] These different senses of the word are not strictly speaking different kinds of piety, but altogether they make up what it means to be *pius*. Conversely, someone who is *impius* is neglectful of these duties. They are ungrateful, irreverent, and unpatriotic—in short, un-Roman.

In stating the problem of the *De div.*, Cicero is repeatedly emphasizing the important connections between Roman self-image, Roman religion, and the legitimacy of the Roman political system. This is all the more significant in that this political system was in considerable turmoil during the course of Cicero's lifetime—and most especially while he was working on the *De div.*—and Cicero was no idle bystander in that turmoil. He was one of the most important players in the attempt to maintain and restore the failing republic. Caesar's dictatorship was brutally ended by senatorial daggers right in the middle of Cicero's writing of the *De div.*, and although Cicero had no part in the actual assassination (having been left out of the highly secret planning by the more action-oriented conspirators), he was still front and center in the various political machinations surrounding not just the death of Caesar, but in what would turn out to be the death of

28. *Leg.* I.31.
29. Linderski 1986, 330.
30. Marius was the other (Szemler 1982, 2316).
31. E.g., *De div.* I.28, I.29, I.105. Marcus himself does something similar at II.75.
32. Cicero, *Rep.* VI.12.

the republic itself. Brutus, for example, flagged Cicero's centrality when he raised his dagger, still dripping hot with Caesar's blood in the Forum, and publicly saluted Cicero by name.[33] Cicero's key involvement in these major events would eventually get him killed at the hands of some of Mark Antony's thugs. The tools of Cicero's political power, the tongue that had delivered his powerful speeches and the hands that had penned his books, were nailed up on public display in the Forum as a gruesome warning to others. This gory end would come just a year and a half after Cicero had finished writing the *De div.*

Nature and the Legitimation of the Republic

The point is increasingly underscored by modern scholars that we cannot understand the *De div.* without situating it in this politico-religious context.[34] Indeed, the safety of the republic is one of the important themes that recurs in much of Cicero's work, both philosophical and rhetorical. Moreover, there are important thematic tie-ins that link the *De div.* to Cicero's earlier pair of more explicitly political dialogues, the *Republic* and the never-finished *Laws*.[35] In particular, the *Laws* goes to some lengths to situate the practices of auspicy at the heart of the Roman political and legal system, not just functionally, but as a central core of its justification and legitimacy. As the argument in the *Laws* gets underway, Marcus is given a task by his interlocutors Atticus and Quintus to outline the legal system of an ideal state, which in the event turns out to be a polished-up version of the Roman republican system. *Et recte quidem,* "and a well chosen subject!" says Marcus, for,

> nam sic habetote, nullo in genere disputandi honestiora patefieri, quid sit homini a natura tributum, quantam vim rerum optimarum mens humana contineat, cuius muneris colendi efficiendique causa nati et in lucem editi simus, quae sit coniunctio hominum,[36] quae naturalis societas inter ipsos; his enim explicatis fons legum et iuris inveniri potest.[37]

33. *Phil.* 2.28.

34. Linderski 1982, 1986; Krostenko 2000; Kany-Turpin 2004.

35. Indeed, there is speculation that Cicero was still polishing the *Laws* around the time of writing the *De div.* or just before. See MacKendrick 1989, 77 and 330n1.

36. Powell 2006 adds *cum dis,* "with the gods," here, which would bring this sentence even more closely in line with *Leg.* I.23, which I discuss in the next paragraph, but the sentence still makes sense without the addendum. Note also the similarity of language with *De div.* II.148, where *religio* is said to be *iuncta cum cognitatione naturae.*

37. *Leg.* I.16.

> You should know that no other kind of discussion brings nobler things to light: what gifts are given to man by nature, how much ability of the greatest kind the human mind contains, what office [*munus*] we have been born and "brought into the light" to strive for and to perform, what the unity of men is, what the natural community is between them—only when this is clarified can the origin of law and justice be discovered.

Here Cicero makes it abundantly clear that the starting point of any understanding of law and justice must begin with *nature.* The very origins, the fount of law and justice, are rooted in the society between people, a society that is itself "by nature," and the ability to reason this out is itself also given to people by nature. (When Cicero refers to the "natural society between men" he does not mean what Hobbes would later call "the state of nature"—nasty, brutish, and short-lived—that people are to be lifted out of by the imposition of laws. Rather Cicero thinks that the right and just society between people is a product of nature itself, and the right way of doing things is to be found in that nature.) This means that the appeal to nature is simultaneously an appeal to society and, for Cicero, political duty within that society. The nature of justice, *natura iuris,* is to be found in the nature of man, *hominis natura,* and law is the highest rationality, innate in nature herself, that governs the cosmos, a point that Cicero emphasizes by repeating it again and again in the *Laws*:

> lex est ratio summa insita in natura.[38]

> Law is the highest reason, implanted in nature.

> nihil est profecto praestabilius quam plane intellegi nos ad iustitiam esse natos, neque opinione sed natura constitutum esse ius.[39]

> Surely nothing is better than to know clearly that we are born for justice, and that law is constituted by nature, not by opinion.

> igitur video sapientissimorum fuisse sententiam, legem neque hominum ingeniis excogitatam nec scitum aliquod esse populorum, sed aeternum

38. *Leg.* I.18.
39. *Leg.* I.28.

quiddam, quod universum mundum regeret imperandi prohibendique sapientia. ita principem legem illam et ultimam mentem esse dicebant omnia ratione aut cogentis aut vetantis dei.[40]

So I perceive that it has been the opinion of the wisest that law has not been invented by the minds of men nor is it some kind of decree made by peoples, but something eternal, which rules the whole cosmos by the wisdom of its commands and prohibitions. Thus they say that this first and final law is the mind of god, compelling or forbidding everything by means of reason.

erat enim ratio profecta a rerum natura et ad recte faciendum impellens et a delicto avocans, quae non tum denique incipit lex esse, cum scripta est, sed tum, cum orta est; orta autem est simul cum mente divina.[41]

For reason existed (already), effected by the nature of things, and impelling the doing of right and avoidance of wrong. It did not finally become law when it was written down, but when it came into being, and its coming-into-being was simultaneous with the divine mind.

ergo est lex iustorum iniustorumque distinctio ad illam antiquissimam et rerum omnium principem expressa naturam, ad quam leges hominum diriguntur.[42]

Thus law is the discrimination between just and unjust actions, a pressing from that most ancient and eminent of all things, nature, and upon which all the laws of men are ordered.

gaudeo nostra iura ad naturam accommodari maiorumque sapientia admodum delector.[43]

I am glad that our laws are in agreement with nature, and I am thoroughly delighted at the wisdom of our ancestors.

40. *Leg.* II.8.
41. *Leg.* II.10.
42. *Leg.* II.13.
43. *Leg.* II.62.

Because of this conflation of law, reason, and divine governance, and because reason is shared by humans and the gods, we get an important expansion of the meaning of "society": *quae [sc. recta ratio] cum sit lex, lege quoque consociati homines cum dis putandi sumus,* "because right reason *is what law is,* we must conclude that it is by law that men are in community with the gods,"[44] which community is explicitly referred to as *civitas,* citizenship. And so, when Cicero finally comes to laying out the details of the specific laws of the ideal state, we find the mapping out of the duties of people to gods as the first order of business. Not just any gods, but public gods, for the public good. Thus at the outset, Cicero establishes not the existence of the gods, for he thinks that is a given, but the parameters and responsibilities of the state religion, and he roots this in a conception of law as given in nature itself.

The idea of public gods serves to remind us that the Roman political system was never just what we would call political, but was always politico-religious.[45] All Roman magistrates were both political and cultic functionaries, and priestly offices were simultaneously public. Of these priestly offices, Cicero tells us, there are three classes: one responsible for ceremonies and sacrifices, and the other two for various types of divination.[46] The diviners are subdivided into those who interpret prophesies for "the senate and the people" (a formulaic phrase meaning something like "our country"),[47] and the interpreters of Jupiter Optimus Maximus, who are the augural college.

Decisions on warfare, public safety, food supply, diplomacy, foreign affairs, and in general all functions of government and state are to be put to the various priests for approval,[48] and (in Cicero's ideal formulation) the decisions of the gods to be obeyed on pain of death. This is not a conflation of what we would think of as "church and state" so much as one of ritual and governance, and its justification is rooted in a concept of the natural consociety of humans and gods that is embodied in the state. So, too, from the *De div.* it is clear that divination is at the heart of the state's very legitimacy.

44. *Leg.* I.23.

45. A good summary can be found in North 1986.

46. *Leg.* II.20–21.

47. The Latin phrase *senatus populusque romanus* was abbreviated to *SPQR* and became something like an official emblem of the empire.

48. The technicalities of how to put the questions and when and where the answers could be announced are discussed in Linderski 1982, 1986.

A Ciceronian Contradiction?

That Cicero's later work is meant to pick up unfinished political business from the *Laws* is further seen on the one hand in the frequent recycling of divinatory stories from the *Laws* in both the *De div.* and the *De natura deorum*, and on the other hand in Cicero's use of a surprising pair of parallel—but completely antithetical—statements. Compare these two assertions, the first in the mouth of Marcus in the *Laws* (II.32), the second in the mouth of Marcus in the *De div.* (II.74):

> *divinationem esse sentio*
>
> I believe in divination.
>
> *divinationem nego*
>
> I deny divination.

The most common line of resolution has been to point to the dates of the two works: the *Laws* was written somewhere in the late 50s BC, the *De div.* in 44. So one could argue for a development in Cicero's thinking, from the piety of the earlier work to the scepticism of the later.[49] Perhaps as the political situation shifted, his position moved in response.[50] Alternatively, we could ask whether the two works were intended for very different audiences: the one public and therefore cautious, the other privately circulated and therefore freer.[51] Other lines of interpretation take a deliberately nonchronological tack. One sees Cicero as adopting different epistemological stances, and so working with different epistemological standards, in the speeches *De domo sua* and *De haruspicum responso* on the one hand, and in *De divinatione* on the other.[52] The argument is that Cicero does believe in portents in the speeches, when he deliberately takes a "religious attitude," and he rejects portents in *De divinatione*, where he takes a "philosophical attitude." These attitudes are then seen as akin to epistemological

49. Guillaumont 1984; Schofield 1986.

50. Jaeger 1910; Krostenko 2000.

51. Pease 1979, 13, argues that the independence of thought in the *De div.* may imply that it was intended only for a close circle of friends rather than for general publication, although I am hard-pressed to know what that might mean given methods of book distribution in first-century Rome, which depended on circles of friends in any case. See Murphy 1998.

52. Goar 1968; Rasmussen 1998.

stances, where standards of truth and validity are to be judged solely from within the stance itself, and so philosophical considerations cannot serve as arguments or objections under the religious attitude, and vice versa. Other scholars have been suspicious of such "brain balkanization."[53] For her part, Mary Beard questions whether the *De div.* really comes to any conclusion at all, and more importantly whether "belief" and "scepticism" as epistemological positions with respect to religion are really contextually appropriate when reading Roman philosophers.[54] The first question is particularly to the point with Cicero, whose use of the dialogue form, especially in his later works, increasingly experiments with open-endedness and with giving expanding room to the play of arguments rather than trying to force the reader to a specific conclusion.[55] As has been pointed out by recent commentators, neither of the endings of the *De natura deorum* and *De div.* says much more than that the arguments on both sides of the respective debates were good, and that an academic sceptic should ultimately suspend judgment. In the *De natura deorum,* Cicero ends by telling us that he thought the Stoic argument in the dialogue the better, but that is tempered by his Epicurean character, Velleius, siding with the sceptics. Moreover, it is not that Balbus the Stoic "won" the debate as though winning a contest; it is that Marcus thinks Balbus' argument "tended more nearly to a resemblance of truth," a very careful and cautious phrasing.[56] At the end of the slightly later *De div.*, Cicero states the Academic sceptic's position even more clearly: he has put forth the best arguments he could for both sides, and he will leave the final judgment in the reader's hands, *iudicium audientium relinquere integrum ac liberum.*

So what, then, do we make of his explicit, and explicitly opposed, statements of belief and disbelief? A clue to their resolution lies in the fact that right after making each of the opposed statements, Cicero does exactly the same thing in both dialogues: he turns to a then recent debate—within the augural college itself—on the status of divination.

The participants in the debate in the augural college are Appius Claudius Pulcher and Gaius Claudius Marcellus. Appius Pulcher was the sometime scandalously corrupt politician and brother of Cicero's long-time enemy Publius Clodius Pulcher,[57] and he was the author of a treatise on divination

53. See, e.g., Veyne 1983; Feeney 1998.

54. Beard 1986. Contrast Takács 1995, 12f.

55. See, e.g., Fox 2007; Schofield 2008; Goldhill 2008.

56. *ad veritatis similitudinem videretur esse propensior. De nat. deor.* III.95.

57. The change in spelling from the more usual Claudius to the more rustic Clodius was a pretence of P. Clodius Pulcher's, adopted as part of his populist political agenda.

dedicated to Cicero (divination was clearly a topic of fairly wide upper-class discussion in the first century BC).[58] Marcellus was the father of the consul of the same name for 50 BC.[59]

What is interesting is that Cicero's two position statements, denying divination in *De div.* and affirming it in the *Laws*, are each immediately situated with respect to the debate between Pulcher and Marcellus, the one claiming divination to be true knowledge (*scientia*) and the other claiming that it has political expediency:

From the *Laws*:

> ita neque illi [sc. C. Marcello][60] adsentior qui hanc scientiam negat umquam in nostro collegio fuisse, neque illi [sc. Ap. Claudio], qui esse etiam nunc putat; quae mihi videtur apud maiores fuisse duplex, ut ad rei publicae tempus non numquam, ad agendi consilium saepissime pertineret.[61]

> Thus neither do I agree with him [Claudius Marcellus] who denies this knowledge was ever had in our college, nor with him [Appius Claudius] who thinks it exists even now. To me it appears that it was twofold among our ancestors: that it aided the republic on occasion over the years, but most often it applied as counsel for administration.

From *De divinatione*:

> equidem adsentior C. Marcello potius quam Appio Claudio, qui ambo mei collegae fuerunt, existimoque ius augurum etsi divinationis opinione principio constitutum sit tamen postea rei p. causa conservatum ac retentum.[62]

> For my part, I agree more with C. Marcellus than with Appius Claudius, both of whom were my colleagues, and I think that the augural law, although at first it was constituted because of a belief in divination, afterward it was conserved and retained for the sake of the republic.

58. Treated at some length by Varro, P. Nigidius Figulus, A. Caecina, and L. Caesar, as well as Cicero and A. Pulcher, among others. For a detailed list and its sources, see Krostenko 2000, 375n55. For the century after this, see Neri 1982.

59. Gundel 1975.

60. For the identifications here, compare Atticus' characterization of their respective positions at *Leg.* II.32.

61. *Leg.* II.33.

62. *De div.* II.75.

In neither passage does Cicero say that he himself thinks divination is simply true or false. He says that Marcellus claims that divination was maintained in the augural college, from its inception, for the sake of the *res publica,* and that Appius Claudius claims that divination is maintained in the augural college because it is true.[63] Marcellus' position (or one just like it) was for a long time taken by modern historians to be symptomatic of a widespread religious cynicism on the part of the Roman nobility, signaled not only by the apparent scepticism of the *De div.*, but also by scandalous behavior such as P. Clodius Pulcher's sneaking into the rites of the Bona Dea, where men were strictly forbidden, while dressed as a woman.[64] Recent studies have to some extent moved away from a belief/disbelief dichotomy, but sceptical readings of various sorts do still see the light of day. Krostenko, for example, thinks that Cicero is arguing in the *De div.* that neither belief nor scepticism is adequate, and instead he attempts to paint a divination that "is purely formal and symbolic,"[65] but I do not see how this avoids the sceptical reading. This track seems doubly determined by Krostenko's calling divination a "useful tool of social control" and his reading of a deliberate "noble lie" in Cicero to keep the lower classes in check.[66] By contrast, I argue that the phrase "for the sake of the republic" need not be taken as a statement of aristocratic deception of their less educated countrymen. Instead, it should be read as a genuine attempt to ground a desperately faltering but—as Cicero saw it—a very honorable and worthy political system. Divination is not a noble lie for Cicero. It is legitimate, but that legitimacy lies in a piety—a sense of duty—that is political. In the *Laws*, it is explicitly and repeatedly made clear that this means "political" in that large Roman sense, the sense that is mindful of our consociety with the gods, and that gains its legitimacy through its grounding in nature.

Knowledge of Nature and Virtuous Action

There are powerful gestures in this same direction in the *De div.* as well. Remembering that in the dialogue immediately preceding the *De div.*, the *De natura deorum,* which was meant to have been read first if Cicero's opening and closing remarks in the *De div.* mean anything, the gods emerge as most

63. Quintus argues against the position of Claudius Marcellus (though Marcellus is not mentioned by name) at *De div.* I.105–7.

64. On cynicism, see the comments in Beard 1986, 33.

65. Krostenko 2000, 376.

66. Krostenko reads too much into Cicero's words at II.70 (2000, 377), notwithstanding the strong classism Krostenko argues for in the dialogue, which I would argue is likewise overplayed.

plausibly true and real, and their relationship to humanity strongly benevolent. Close attention to Marcus' concluding monologue in the *De div.* further strengthens this point, and brings us back once again to nature:

> nam, ut vere loquamur, superstitio fusa per gentis oppressit omnium fere animos atque hominum imbecillitatem occupavit. quod et in iis libris dictum est, qui sunt de natura deorum, et hac disputatione id maxime egimus. multum enim et nobismet ipsis et nostris profuturi videbamur si eam funditus sustulissemus. nec vero (id enim diligenter intellegi volo) superstitione tollenda religio tollitur. nam et maiorum instituta tueri sacris caerimoniisque retinendis sapientis est, et esse praestantem aliquam aeternamque naturam, et eam suspiciendam admirandamque hominum generi pulchritudo mundi ordoque rerum caelestium cogit confiteri. quam ob rem ut religio propaganda etiam est, quae est iuncta cum cognitione naturae, sic superstitionis stirpes omnes eligendae.[67]

> To speak truly, superstition, widespread among peoples, has borne down the minds of all men and preyed on their weaknesses. This was said in my book *De natura deorum,* and I have targeted it most especially in the present dialogue. I thought I could do a lot for all of us if I were able to overthrow it at the foundations. But emphatically—and I want this understood clearly—the overthrowing of superstition is not the overthrowing of *religio.* Indeed, it is for the wise to protect the institutions of our ancestors, maintaining their rites and their ceremonies.

Then we get Marcus' powerful move to *nature*:

> So also the beauty of the cosmos and the order of its phenomena compel us to acknowledge the existence of an excellent and eternal nature that is honored and admired by the race of men. Thus just as *religio,* which is part and parcel with a knowledge of nature, is to be cultivated, so all the roots of superstition must be rooted out.

The appeal to the importance of maintaining the *instituta maiorum,* the institutions (but also habits and manners) of the ancestors, is rhetorically a powerful one, tying very strongly into Roman historical self-conception. It may also be meant to call to mind not only public ritual, but also private,

67. *De div.* II.148–49.

household ritual, where rites and traditions surrounding the Penates, Lares, and the *genius* (a kind of spirit) of the head of the household, the *paterfamilias,* are passed from parents to children within the confines of the family home. Here too, though, the state is never far away, since the empire as a whole was seen to be protected by Penates and *Lares praestites,* and there was also a *genius* of the Roman people as a whole, in a kind of extension of the rituals of the private home to the state as a whole.[68]

But the real crux of Cicero's argument has to do with how nature is tied to *religio.* We shall see in the next chapter how nature was taken by many authors in antiquity to underscore ethics more broadly, but for now we need to focus on Cicero's reading of *religio* and its cognate virtue, *pietas.* In this passage, *religio* is not linked to nature directly, but instead to the *knowledge* of nature, *cognitio naturae.* The difference is an important one, and it is where Cicero finds a special place for the *sapiens,* the wise man, in the upholding of the rites and ceremonies of the ancestors. It is not that nature forces *religio* on humanity as part of our constitution, or that *pietas* is simply built into human nature in the same way that our desire to live together in groups is said to be given by nature in the *Republic* and the *Laws.* It is instead a claim that *religio* emerges from the learned and careful contemplation of nature, and it is from this knowledge that the *sapiens* gains his political and moral authority, as well as where he situates the grounds for the obligations he owes to the state.

It may also be worth remarking that this reading of Cicero accords in important respects with a recent reading of Varro's *Divine Antiquities* by Peter Van Nuffelen, in which Cicero's friend and codedicatee Marcus Terentius Varro (writing a major theological work, likely just a little before Cicero's *De natura deorum*) is seen to be rooting his conception and characterization of the traditional Roman pantheon in a knowledge of the physical workings of the cosmos.[69] Unfortunately the *Antiquities* is too fragmentary for us to say much more on this point, but Varro's emphases on and connections between *theologia naturalis* and *theologia civilis,* as well as his known worry about *superstitio,* are telling. We should not, though, overstate the similarities. Where Cicero's *religio* is rooted in a knowledge of nature that is possessed by wise men in the present, Varro's grounding seems to be historical, in that the rites, deities, and institutions of Roman religion preserve an ancient knowledge of the world that was allegorically encoded by wise

68. See Orr 1978.

69. Van Nuffelen 2010. See also Ando 2008.

men in the mists of time. So, too, there are hints that Varro is quite happy to see some false practices and beliefs promoted disingenuously because they are "useful" for the populace.[70]

As the opening of the *De natura deorum* shows, it would be hasty to read into Cicero a similar aristocratic cynicism.[71] There (perhaps in direct response to Varro?) Marcus opens the dialogue by saying that *pietas*, duty, cannot be just for show:

> in specie autem fictae simulationis sicut reliquae virtutes item pietas inesse non potest; cum qua simul sanctitatem et religionem tolli necesse est, quibus sublatis perturbatio vitae sequitur et magna confusio; atque haut scio an pietate adversus deos sublata fides etiam et societas generis humani et una excellentissima virtus iustitia tollatur.[72]

> Just as for the other virtues, it is not possible for *pietas* to consist in mere show or feigned pretence, since both *sanctitas* and *religio* would necessarily be overthrown at the same time, and both disorder in our way of life as well as great disturbance would follow such an overthrow. It seems to me that with the overthrow of *pietas* to the gods would come the overthrow of trust, the society of the race of men, and that most excellent virtue, justice.

The language and structure of this passage mirror to a large extent those of Marcus' speech at the end of the *De div.*, quoted just above.[73] The performance of the rites at the core of the Roman state cannot, in either passage, consist just in going through the motions. Marcus says in the *De natura deorum* that the act of performance assumes at the outset the existence of the gods, without whom everything would fall apart.[74] As he situates it here and in the *Laws*, *religio* is at the core of the functioning of the state—and he goes so far as to put this very point in the mouth of even his sceptical character, Cotta, in the *De natura deorum*.[75] Looking to the *De div.*, we see an insistence that there is a dark shadow to *religio* that Cicero is very

70. Augustine, *De civ. D.* 4.31.

71. *Pace* Krostenko 2000, I very much doubt Cicero would or could claim *sapientia* to be the province of the nobility in general.

72. *De nat. deor.* I.3.

73. Note especially the various "overthrowings" and the worries about their combinations and snowball effects.

74. *De nat. deor.* I.2.

75. *De nat. deor.* III.5.

concerned about bracketing away: *superstitio.*[76] Still, in all three of the dialogues, the *De div.*, the *De natura deorum*, and the *Laws*, Cicero worries repeatedly about safeguarding the state's religious apparatus.[77] We could see the worries about superstition in the *De div.* as being akin to the worries about "private gods," which undermine the functioning of the state and lead to a *confusio religionum* in the *Laws.*[78] In any case, what emerges repeatedly is an insistence that the maintenance of the official cult is absolutely central not just to the maintenance of the state as it stands, but—and here Cicero pushes his view of Romanness to the fore—to the maintenance of justice itself, and of all human society. And it is precisely in this maintenance of human society that virtue approaches most nearly to divinity: *neque enim est ulla res, in qua propius ad deorum numen virtus accedat humana, quam civitatis aut condere novas aut conservare iam conditas,*[79] "Nor is there any other matter in which human virtue comes nearer to the majesty of the gods than in the building of new cities, or the conservation of those now standing."

But how does Cicero bring his reader to acquiesce to these sweeping claims? In the *De natura deorum,* as in the *Laws,* Cicero opens with a version of "we hold these truths to be self evident," and then he buttresses the position considerably through the mouths of Balbus the Stoic and of Marcus himself. But at the end of the *De div.* Cicero takes a different turn. There, as in the earlier *Laws,* we see that *religio* comes bundled with an understanding of nature (personified and divine it would seem, given his wording). Curiously, though, we come to recognize this nature through seeing the order and beauty of the cosmos, thus through the senses, which were often seen as unreliable in antiquity. Here we open up one of the larger questions of the present book, the question of how observation is used in arguments about nature, what criticisms it faces, and how various authors try to constrain or control the problems posed by observation.

In the case of the argument of the *De div.*, we find a tension between Quintus' unsuccessful enumerative argument, citing observation after observation and case after case, and Cicero's appeal to the observation of beauty and order in the cosmos as the key source from which human knowledge of a divinized Nature derives. So how and why does Marcus reject Quintus' observations while maintaining the importance of his own?

76. A line (δεισιδαιμονία) Plutarch warns us not to cross as well. *De Iside et Osiride* 355D.

77. Krostenko quite rightly points to Caesar's use of religious symbols and innovation as an important point of objection for Cicero in the *De div.* (2000, 380, 384).

78. *Leg.* II.25.

79. *Rep.* I.12.

Fabulae versus Learned Observation

It has often been remarked that Quintus' argument in book 1 of the *De div.* is a congeries of quite disconnected stories. We see Tarquinius Priscus cutting a whetstone in half with a razor; Attus Navius, in search of a suitable offering to the gods, using the augural staff to find grapes of a "marvelous size" in his vineyard; and the senior Tiberius Gracchus' dilemma where, depending on which of two captured snakes he released, it would result in his own or his wife's death. We are told of prophetic dreams: Dionysius the Tyrant's mother giving birth to a Satyr, Tarquinius Superbus being butted by a goat and seeing the sun reverse its course, the Persian king Cyrus chasing but never catching the sun at his feet, Simonides being warned by a ghost not to board a ship. Quintus relates tales of prophecy and frenzied visions. Ships are sunk, battles are lost, kingdoms gained and ruined, all foretold in one way or another. Not only do we see ghosts and visions jumbled up with mindreading and marvels and divine-intervention stories, we also get ritual divinations (sacred chickens, extispicy) stood up beside random ominous events (lightning, birds, snakes, and bugs).

But there is a strong theoretical principle underlying Quintus' argument.[80] He insists repeatedly that he can provide a justification for divination that does not require an appeal to causes. So in his response to Quintus in book 2, Marcus zooms in on this claim, insisting instead on the necessity of a causal account. At this level, the debate is no longer simply about divination, but about the acceptability of particular types of argument and evidence. Thus Quintus:

> quarum quidem rerum eventa magis arbitror quam causas quaeri oportere.[81]

> But of these things I conclude it more proper to look at results than at causes.

> atque his rerum praesensionibus prognostica tua referta sunt. quis igitur elicere causas praesensionum potest?[82]

80. I do not follow Krostenko in seeing Quintus' "disordered" argument as an attempt to distract the audience from the main issue (2000, 373).

81. *De div.* I.12.

82. *De div.* I.13.

> And your [Cicero's] *Prognostics* is full of presentiments of these things, but who can ascertain the causes of these presentiments?
>
> videmus haec signa numquam fere mentientia nec tamen cur ita fiat videmus.[83]
>
> Almost never do we see these signs to be false, and yet we do not know why it so happens.
>
> non quaero cur, quoniam quid eveniat intellego.[84]
>
> I do not ask *why*, because I understand *what* happens.
>
> hoc sum contentus, quod etiamsi cur quidque fiat ignorem, quid fiat intellego.[85]
>
> I am content being ignorant of why it happens, but *that* it happens, I understand.
>
> similiter quid fissum in extis quid fibra valeat accipio; quae causa sit nescio.[86]
>
> Likewise, what the cleft in the liver, or what the entrails mean, I accept. What the cause is, I do not know.

And more. That Quintus sees the avoidance of causation as a positive strength of his argument is shown most clearly at *De div.* I.85: *nec vero quicquam aliud affertur, cur ea quae dico divinandi genera nulla sint, nisi quod difficile dictu videtur quae cuiusque divinationis ratio quae causa sit,* "Really, no other argument is offered for why the kinds of divination I am talking about are nonexistent, other than pointing out that it is difficult to give a reason, a cause, of the various kinds of divination." He continues: *cur fiat quidque, quaeris; recte omnino, sed non nunc id agitur: fiat necne fiat, id quaeritur,*[87] "You ask why everything happens, and quite rightly so,

83. *De div.* I.15.
84. *De div.* I.15.
85. *De div.* I.16.
86. *De div.* I.16.
87. *De div.* I.86.

but that is not now under discussion. Does it happen or does it not happen? That is at issue."[88]

So, too, Marcus wastes little time cutting to the methodological question and insisting that causes are needed if Quintus' stories are to be convincing. In his rejection of Quintus' argument, Marcus uses language identical to that in the opening monologue of the *De div.*, where he had pointed to a tension between proof by argument and proof by example: *atque haec, ut ego arbitror, veteres rerum magis eventis moniti quam ratione docti probaverunt,*[89] "This is what I conclude: that the ancients were more convinced in these matters by the results of divination than they were by learned argument." When he comes to aim the full force of his argument in book 2 (after "skirmishing with light arms"), he says: *argumentis et rationibus oportet quare quidque ita sit docere, non eventis, iis praesertim quibus mihi liceat non credere,*[90] "It behooves you to show by arguments and reasons, not by results, why these things are as they are, and especially not by results that I am free to disbelieve." Marcus expands on this point in the (alas, fragmentary) sequel to the *De div.*, the *De fato,* when he brings up an argument that divination would be impossible, even for Apollo, unless the events predicted were the result of a necessary causal chain.[91] But the worst part, as he makes abundantly clear, is that the stories Quintus is marshaling are just too incredible. In fact, Marcus uses the same word to discredit them as we will see Seneca using to discredit all the unworthy stories about the Campania earthquake in chapter 4. "Surely," says Marcus, "you don't think I'll believe *fabulae*?"[92]

This contrast between *fabulae* and some better sort of observation is important for Cicero, for it lets him employ certain kinds of sensory evidence in his larger argument, but at a much more refined level than Quintus' ragbag of tales. Not only were Quintus' stories pointing in many different directions at once, they also came from many different kinds of sources, from poets to historians to simple gossip. The examples are a kind of experiential flotsam, and are individually passed on by random reporters—reporters who are not sufficiently philosophically rigorous to really know how to critically analyze what they are seeing. Contrast the kind of seeing that shows Cicero, at the end of the *De divinatione,* that the cosmos is so beautifully ordered,

88. This is in spite of the fact that Quintus really does believe that there is a cause: *est enim vis et natura* (*De div.* I.12). Compare also I.110f., I.118, I.125.

89. *De div.* I.5.

90. *De div.* II.27.

91. *De fato* 33.

92. *De div.* II.113: *num igitur me cogis etiam fabulis credere?*

that is a different kind of seeing altogether. Look back at the closing monologue to the *De div.*:

> et esse praestantem aliquam aeternamque naturam, et eam suspiciendam admirandamque hominum generi pulchritudo mundi ordoque rerum caelestium cogit confiteri.[93]

> So also the beauty of the cosmos and the order of its phenomena compel us to acknowledge the existence of an excellent and eternal nature that is to be honored and admired by the race of men.

This kind of seeing is in the hands of the *sapiens.* This seeing is abstracted from bare sensory experiences insofar as it is not any one particular seeing-event that proves the existence of this excellent and eternal nature. It is instead the seeing of *beauty* in the cosmos, its order, which is a higher level of observation altogether. From that seeing we get a personified and seemingly divinized nature, to be honored and admired by people. Only when we come to know nature—perhaps better, Nature—can we fully understand *religio,* our duty to the gods, and the core of the best possible state.

Conclusion

Thus careful observation of higher-order aspects of nature (its beauty, its order) leads inevitably to proper ethical behavior, both between people, and between people and the gods. The trained and moderated eye, then, sees in the world around us the foundations of a particular political philosophy. We will broaden the implications of this connection between nature and normative human action in the next chapters. What I have hoped to establish in this chapter is the situatedness of the Roman discourse on nature within a larger intellectual framework, a framework that is importantly both theological and, for Cicero, political and ethical. The way he moves so easily and elegantly between nature, *religio,* and ethical duty to the state is very much a product of his historical time and place, both in terms of his external circumstances (fatal crises in the republic he is trying desperately to save) and in terms of the larger intellectual culture in which he finds himself. It is clear from the introductions to virtually every philosophical text he writes during this, his most prolific period, from the *De finibus* and *Academica* of 45 to his final philosophical work, the *De officiis* of late 44,

93. *De div.* II.148.

that Cicero is acutely conscious of how deeply his philosophy is indebted to its Greek predecessors. But he emphasizes again and again that he wants his reader to see that he is building something unique and new upon those foundations, and as he stresses repeatedly, it is important that he is doing so in Latin and for Romans. This forefronting of his own philosophy as Roman, ubiquitous as it is in his later works, is apparent even in his philosophical efforts from the 50s BC, as when Scipio remarks in the early part of the *Republic* that he is discoursing not as a man who has studied Greek philosophy (although he admits to having done so) but as a Roman, *ut unum e togatis,* who has learned more from *domesticis praeceptis* than from foreign books.[94] Here *domesticis praeceptis* carries nuances difficult to translate, referring to home-grown knowledge (with a hint of folksiness to it), and the word *domesticus* emphasizing, just as it does in English, house and family at the same time as it can be taken to exclude foreign learning (the larger sense of "domestic" as in "domestic policy" and "domestic markets").

By stepping back from subjects that may be more easily and straightforwardly understood as "science," and focusing on a set of arguments about the importance and grounding of Roman divination, we have seen not only how nature is understood within a theological context, but also how integrated the knowledge of nature is with these other culturally important discourses. More particularly, we saw how, at a period when Rome itself is at stake, Cicero tries to find a foundation in nature not only for the Roman political system, but also more broadly for an ideal of Roman virtue that would justify this system as a whole (and not coincidentally his own position as an intellectual within that system).

94. *Rep.* I.36.

CHAPTER THREE

Law in Nature, Nature in Law

We have seen how nature was taken by Cicero as the foundation of a legal and ethical framework, closely tied to his political concerns in the last days of the republic. In this chapter we further open up the question of the relationships between law and nature, finding that the paths of influence are bidirectional in Roman sources, nature grounding ideas of law as well as law grounding ideas of nature. The question then arises of how law is to be understood in this context, given that "laws of nature" as a way of understanding natural regularities are usually supposed in the history of the sciences to be an early modern invention.

Knowledge about nature—at least, knowledge of a sophisticated kind—is these days often glossed as knowledge about something called "the laws of nature." But what do we mean by that phrase? We often talk of nature as being *lawlike,* but don't often reflect on what exactly law-likeness is supposed to imply. After all, it is a bit of an odd juxtaposition, law and nature. Human rules and regulations meet the (almost by definition) doer-of-its-own-thing. We legislate in order to regulate human behavior, but nature does as it will, not as we will it to do. The disjunction here is perhaps best captured by Gerry Mooney, in his brilliant send-up of a mid-1970s American highway-safety campaign that admonished drivers with the dramatically voiced tagline "Buckle up. It isn't just a good idea. It's the law" (see fig. 3.1).

The oddness, the slight conceptual disorientation that emerges from Mooney's juxtaposition does not dissipate readily. And this is strange, for law and nature are long-time companions. In the last thirty years it has become clear that Mesopotamian science, the earliest we have in the his-

A version of this chapter was published as Lehoux 2006a.

Fig. 3.1 Gerry Mooney on Laws and Nature. Gravity Poster design ©2005 Gerry Mooney.

torical record, has important legalistic forms and formulae.[1] Greek science emerges in a rhetorical context that is never very far from judicial practice, and even early Greek cosmological models make

1. Larsen 1987; Jeyes 1989; Ritter 2005; Lehoux 2003b; Rochberg 2004.

common use of judicial and political metaphors.[2] Nevertheless, modern historians have been hesitant to see these conceptions as laws of nature in our modern sense—or perhaps I should put that phrase "our modern sense" in scare quotes, since there is much disagreement between philosophers about what the phrase *laws of nature* may mean, or even whether it means anything at all.

When we look to the ancient sources to see exactly how law and nature come together, we find that something interesting is happening, and it is happening predominately in Latin sources beginning in about the first century BC. Roman authors, it turns out, talk quite unambiguously of "the laws of nature." Now the question becomes: What do they mean by it?

To answer this question, we need to explore both what they mean by *law* and what they mean by *nature,* and then see what happens when the two ideas meet in the wild. I argue that there are two directions to this encounter, one where law bears on nature (in something like the "laws of nature" sense we have been discussing), and one where nature bears on law (where human justice is seen to be rooted in nature in some way, as it was for Cicero). Here it is worth introducing a terminological distinction, since we now have two very different ways in which law and nature can interact. For the sake of clarity, I will refer to any societal law or system of morals that is supposed to be grounded in human or universal nature as a *natural law* theory, and use *law of nature* exclusively to refer to "$F = ma$" type statements.[3] An important theme that emerges over the course of this book is that we cannot understand either use, neither the scientific nor the judicial ("ethical," more broadly) in isolation. Indeed, the interrelationships between ideas of law and ideas of nature that come to their fullest fruition in Roman science form one of the most intriguing and powerful frameworks we find in the ancient world for understanding both the natural world and the duties of humans in that world.

Laws of Nature

Roman literature in the first century BC and the first century AD makes fairly common reference to natural processes as being constrained by laws, although some sources employ an otherwise unusual turn of phrase. In place of, but occasionally alongside of, *leges naturae,* the phrase we might expect

2. Lloyd 1966, 1979; Asper 2007.

3. $F = ma$ is the modern notation for Newton's second law (force equals mass times acceleration).

on a naïve translation of our own terminology into Latin, we also find *foedera naturae.*[4] The latter phrase is traditionally translated, following Lewis and Short's *Latin Dictionary,* as "laws of nature," and it is clear from a key passage in Lucretius (mid-first-century BC) that *foedus* (treaty, covenant, stipulation) is being used interchangeably with *lex* (law, rule, regulation, but also like *foedus,* contract, covenant). At the outset of book 5, just before Lucretius turns to his argument that the world was not divinely created and that it is also ultimately subject to destruction, he says:

> Cuius ego ingressus vestigia dum rationes
> persequor ac doceo dictis, quo quaeque creata
> foedere sint, in eo quam sit durare necessum
> nec validas valeant aevi rescindere leges,
> quo genere in primis animi natura reperta est
> nativo primum consistere corpore creta,
> nec posse incolumem magnum durare per aevum.[5]

> Treading in his [Epicurus'] footsteps, I have been following his doctrines and explaining in my verses the necessity for each thing to abide by the law [*foedus*] that governs its creation, and its impotence to rescind the strong laws [*leges*] of time. Most important of all, we have discovered that the soul is fashioned and formed of substance that is subject to birth; and we have found that it cannot endure unscathed through vast eternity.

That the perishing of all things that come to be is one law of nature is reaffirmed several times in the poem.[6] This is not to say that atoms themselves ever perish—that would be impossible—but that the things made up of atoms (rocks, trees, the earth itself) must always eventually dissolve, their atoms becoming mixed and mingled differently, creating other, new entities. Note that this does not mean that we are tending toward any ancient equivalent of thermodynamic "heat death," where eventually everything dissolves into stagnancy, but instead that any and every individual existent thing that is compounded of atoms is, by a law of nature, subject to change, seen as decay into other things.[7]

4. Compare Asmis 2008.

5. Lucretius, *DRN* V.55f. Throughout this chapter I use Martin Ferguson Smith's translation of Lucretius, slightly modified.

6. E.g., Lucretius, *DRN* V.306f. and I.584f.

7. Lucretius, *DRN* V.306.

But the law governing change is not just this long-term process of decay and destruction. In the shorter term, we see that individuals of each species of living thing do just the opposite: they instead grow and develop over time. This, Lucretius says, also happens according to laws:

> Denique iam quoniam generatim reddita finis
> crescendi rebus constat vitamque tenendi,
> et quid quaeque queant per foedera naturai,
> quid porro nequeant, sancitum quando quidem extat,
> nec commutatur quicquam, quin omnia constant
> usque adeo, variae volucres ut in ordine cunctae
> ostendant maculas generalis corpore inesse,
> inmutabilis materiae quoque corpus habere
> debent ni mirum.[8]
>
> Furthermore, since in the case of each species, a fixed limit of growth and tenure of life has been established, and since the powers of each have been ordained by solemn decree in accordance with the laws [*foedera*] of nature, and since, so far from any species being susceptible of variation, birds display on their bodies the distinctive markings of their kind, it is evident that their bodies must consist of unchanging substance.

This passage is fairly busy with legislative metaphors. In addition to the *foedera naturae,* Lucretius also uses Latin legal jargon to tell us that the powers of each thing have been ordained, decreed, or enacted as law (*sancio*). He then goes on just after this excerpt to tell us that this law functions as a kind of ontological boundary stone, policing, as it were, the distinction between what can and cannot be, and limiting each thing to its proper scope, a point he repeats verbatim three more times in the poem.[9]

The claim that the proper development of each species is governed by a law is underscored again later, in book 5:

> propterea quia quae de terris nunc quoque abundant
> herbarum genera ac fruges arbustaque laeta
> non tamen inter se possunt complexa creari,

8. Lucretius, *DRN* I.584f.

9. *quid possit oriri, quid nequeat, finita potestas denique cuique qua nam sit ratione atque alte terminus haerens.* Compare *DRN* I.76–77; V.89–90; VI.65–66.

sed res quaeque suo ritu procedit et omnes
foedere naturae certo discrimina servant.[10]

For the things that even now shoot in profusion from the earth—the various kinds of grasses and crops and exuberant trees—cannot, despite their abundance, be created intermixed: each proceeds in its own manner, and all preserve their distinguishing characteristics in conformity with an immutable law of nature.

So not only do the characteristics of a species carry from parent to offspring according to a law, but also characteristics cannot be picked up in passing from other species around the developing individual. Something, a law, keeps grasses from becoming treelike and trees from becoming grasslike, even though they grow profusely around each other.

Thus far, the growth, development, constancy, and decay of all things, animate and inanimate. But other, more specific or more local natural phenomena can be described as obeying laws:

Quod super est, agere incipiam quo foedere fiat
naturae, lapis hic ut ferrum ducere possit,
quem Magneta vocant patrio de nomine Grai,
Magnetum quia sit patriis in finibus ortus.[11]

Turning to another subject, I will proceed to explain by what law of nature [*foedus naturae*] it comes about that iron can be attracted by that stone which the Greeks call the magnet (after the name of its place of origin, the territory of Magnesia).

Lucretius' answer to magnetism is complex, taking just under two hundred lines to complete, but the point emphasized at the outset is just that the magnetic attraction of iron is explainable by Epicurean atomism (even though atomism's prime motive force is pushing, which would appear to be an unpromising start). More importantly for our purposes, the terminology he uses to describe the phenomenon expressly says that magnets attract iron according to a law. And indeed, Lucretius speaks of causation in general as obeying a *foedus* in a richly nuanced passage that is meant as an attack on a rigid determinist idea of fate (such determinism was characteristic of the

10. Lucretius, *DRN* V.920f.
11. Lucretius, *DRN* VI.906f.

Epicureans' main rivals at Rome, the Stoics). To speak of causation as law-like, Lucretius invokes the famous Epicurean "swerve," an occasional and random change in the trajectory of some atoms that was an essential part of Epicurean physics. As Lucretius explains it, without a swerve all the atoms in the cosmos would have all been moving in one direction at the same speed, with no possibility of interaction. If we now imagine an occasional atom deviating from this uniform shower, we can imagine the chain reaction it precipitates creating the possibility of change, including the eventual creation of matter as we know it. Not only does the swerve have important implications for the possibility of atomic interaction, it also gets around the constrictions of Stoic fatalism. Lucretius argues that because the swerve is an uncaused motion,[12] in effect it breaks the law of strict chain-reaction causation and so allows for free will:

> Denique si semper motus conectitur omnis
> et vetere exoritur <motus> novus ordine certo
> nec declinando faciunt primordia motus
> principium quoddam, quod fati foedera rumpat,
> ex infinito ne causam causa sequatur,
> libera per terras unde haec animantibus exstat,
> unde est haec, inquam, fatis avolsa voluntas,
> per quam progredimur quo ducit quemque voluptas,
> declinamus item motus nec tempore certo
> nec regione loci certa, sed ubi ipsa tulit mens?[13]

> Moreover, if all movements are invariably interlinked, if new movement arises from the old in unalterable succession, if there is no atomic swerve to initiate movement that can annul the laws of fate [*foedera fati*] and prevent the existence of an endless chain of causation, what is the source of this free will possessed by living creatures all over the earth? What, I ask, is the source of this power of will wrested from fate, which enables each of us to advance where pleasure leads us, and to alter our movements not at a fixed time or place but at the direction of our own minds?

"Laws of fate," metaphysical though the phrase may appear at first, really just means the laws of causation generally. This is because for the Stoics,

12. It is Cicero who says that the swerve is uncaused. See *De fato*, 22.
13. Lucretius, *DRN* II.251f.

who had a rigidly deterministic physics (every cause has a specific and nonrandom effect), fate emerges as a byproduct of the rigid chain of causation (the French mathematician Pierre-Simon Laplace would later make a similar point when he said that a being who could accurately track every particle in the universe could predict the future to all eternity). But since that Stoic fate is just the unbreakable chain of natural causation, the *foedera fati,* the "laws of fate" that the swerve breaks are essentially the rules of an excessively causal physics: if cause-and-effect are invariable, then the universe is deterministic from the beginning, and everything that will happen is, to use the ancient terminology, fated. By introducing a minor but highly significant uncaused motion, the Epicureans get a cosmos that is not strictly deterministic even if it is causal overall. The Stoic description of fate-*cum*-physical-causation as lawlike, and its implications for human free will are common coin within the school. So we find the interposition of the physics of stellar influence and the determinism of human action in the first-century AD astronomical poem by the Stoic Manilius, who says that the connections between the stars and human destinies are governed by the *fortunae leges,* the laws of fate.[14] Manilius is here connecting human destinies not just to physics generally, but also to the distant stars, a conclusion shared by many ancients, and one that follows naturally on assumptions of regularity and the interconnectedness of all matter.

In Vergil's *Georgics,* a poem that often deliberately hearkens back to the earlier Lucretius,[15] we find:

> ac prius ignotum ferro quam scindimus aequor,
> ventos et varium caeli praediscere morem
> cura sit ac patrios cultusque habitusque locorum,
> et quid quaeque ferat regio et quid quaeque recuset.
> . . . continuo has leges aeternaque foedera certis
> imposuit natura locis . . .[16]

> But before we till the unknown field with iron let us first be careful to learn the winds and the changeable ways of the weather and the particular cultivation and qualities of the location, and what each region grows and what each refuses. . . . Nature has always imposed these laws [*leges*] and eternal edicts [*foedera*] on particular places . . .

14. Manilius, I.56.
15. See Gale 2000.
16. Vergil, *Georg.* I.50–61.

Here we see Vergil paralleling *leges* and *foedera,* and talking of regional climactic variation and local regularity as being subject to these laws. A generation or so later, Ovid expresses his deep shock at a friend's betrayal by likening it to a breaking of the laws of nature.[17] What he thinks these laws of nature are, can be seen in the dramatic events that he thinks would transpire when such laws fail to function: the sun moving backward, rivers flowing uphill, water producing flame. In the *Metamorphoses,* too, Ovid refers to the law by which the stars move.[18] Indeed, the regular motions of the stars are among the most common loci for laws in nature, as for example in Manilius: *sidera . . . exercent varias naturae lege choreas,* "the stars perform their various dances according to the law of nature."[19] And the movements of the stars are frequently taken as the very paradigm of ordered motion in antiquity.

We can further add Pliny's discussion of the law of nature (*lex naturae*) that governs the regularity of the winds.[20] While this law is not perfectly known, says Pliny, we do have a grasp of it, and so some knowledge of how the winds will move. Cicero had, in the century before Pliny, argued that the whole universe was governed by a divine and eternal law (*lex*), and that all natural processes were constrained or directed by that law.[21] Heinrich Von Staden and Brad Inwood have further discussed at length the ways in which the natural world is quite explicitly described as being lawlike in Celsus and Seneca respectively.[22]

It is clear that there is little hesitation among Latin authors—writing on subjects from physics to physiology to farming—in using the idea of law to characterize strict regularity in nature. That these uses of laws of nature build on and develop older Greek ideas seems clear. Geoffrey Lloyd long ago showed that even in some of the earliest philosophical treatments of nature that we have, social and political analogies find common application, and verbs of "governing" and "steering" are often used as metaphors for how

17. Ovid, *Tristia* I.viii.5.
18. Ovid, *Meta.* XV.1.68f.
19. Manilius, I.670–71.
20. Pliny, *NH* II.xlv.116.
21. Cicero, *Leg.* II.8.
22. Earlier versions of chapter 8 of Inwood 2005 played an important role in my own framing of the present discussion, and Inwood's collection of examples of laws of nature in Seneca offers a powerful adjunct to my argument here. Von Staden 1998 shows the ways in which Celsus's medicine depends on a clear conception of when and how we can talk of the behavior of bodies and disease as lawlike, and alerts us to the question of whether jurisprudential terminology is a necessary component of lawlike accounts of nature.

some overarching causal principle acts in the cosmos.[23] For Lloyd, what distinguishes the governing that the "air" or the "reason" of the philosophers performs from the governing that Zeus performs in contemporary poetic or mythological imagery is the impersonal nature of the physical principle among the philosophers. In fact, though, philosophers are far from universally agreed on this last point—Empedocles' Love and Strife, Anaxagoras' Mind, Heraclitus' Zeus, and Plato's Demiurge all share important characteristics with personal divinities that we should not understate.[24] So, too (as Lloyd acknowledges), it is often difficult to determine the force of a political or social analogy, particularly in the earlier philosophers. Thus when Simplicius, writing in the sixth century AD, reports that the Presocratic philosopher Anaximander characterized the generation and destruction of things as a making of amends and payment of reparations (διδόναι γὰρ αὐτὰ δίκην καὶ τίσιν ἀλλήλοις τῆς ἀδικίας κατὰ τὴν τοῦ χρόνου τάξιν),[25] Simplicius immediately backs away from a literal interpretation by pointing out that Anaximander was being "rather poetic" (ποιητικωτέροις οὕτως ὀνόμασιν αὐτὰ λέγων). Plato, too, struggles in the *Timaeus* with how literally we should understand his images of cosmic rulership, settling on a notion of "likeness" or verisimilitude-to-the-truth (εἰκός), variously calling his account both a *logos* and a *mythos*.[26] Finally, Lloyd showed that the political metaphors in early cosmology were analogous to the uses of biological metaphors in Greek political theories, where the state is likened to a living organism, and to political metaphors in Greek medicine, where the living body is likened to a state.[27] Neither metaphor seems to have been meant literally.

But something more than metaphor seems to be at play in the Roman sources. We have already explored Cicero's conflation of right reason, law, and nature, an idea that seems to owe a considerable debt to Stoic natural-law philosophy. But at the same time Cicero's intellectual context seems to have decoupled these ideas from Stoicism, signaled by the easy agreement Cicero has his interlocutors give to the propositions—interlocutors of all philosophical stripes, including the Epicurean Atticus in the *Laws*. Indeed, as we have seen, the phrase "law of nature" is widespread in Latin, occurring across a wide range of authors with a wide range of philosophical commitments. To say that some cause acts "by necessity" is one thing (and the Greeks were no strangers to such talk), but to have an apparently philo-

23. Lloyd 1966, 210f.
24. See, e.g., Sedley 2007.
25. Anaximander, fr. B1 DK.
26. *Timaeus*, 29c. See Johansen 2004, esp. 48f.
27. Lloyd 1966, 232f., 396f.

sophically diverse community commonly describing a wide range of causal relations as "laws of nature," from magnets to inheritance to seasonal variations, is quite another. And this use is a little surprising, insofar as (a) with one possible exception it does not seem to show up in Greek,[28] and (b) the modern debates on the origins of the idea of laws of nature have put those origins fourteen or fifteen centuries later than what we find in these sources. Thus two questions: What is the intellectual context in Rome that allows for the juxtaposition of law and nature in this new way? And why have these ancient sources been discounted by modern historians? We will see that, in the process of addressing these problems, a third and very difficult question rears its head, in the need to clarify more precisely what we even mean by the phrase "laws of nature" before we can begin to attribute a version of it to anybody, ancient or modern.

Natural Laws

As we look to the ancient philosophical contexts, we notice that juxtapositions of law and nature are not limited to descriptions of natural regularities. In ethics, too, law and nature come together to ground several prominent theories of morality, politics, and jurisprudence, from antiquity to the present day (we have already seen the example of Cicero).[29] Without trying to establish an earliest date or a first author, we can see that the big schools of philosophy that had developed in the Hellenistic period were in large part—many of them in largest part—dedicated to ethics as the primary focus of the school's teaching. Many schools saw their physics and their logic as deeply connected with, and in some cases primarily as instruments in the pursuit of, ethical ends. Even Epicurean physics, whose billiard-ball atomistic cosmos appears to the modern eye as the least teleological of all the ancient physical schools, was always geared toward the promotion of happiness and the elimination of fear as its primary task in both education and investigation.[30] Likewise, Stoic ethics is thoroughly and characteristically grounded in the Stoic conception of nature. I say characteristically here just because

28. The Hippocratic work *On Generation/On the Nature of the Child* opens with the aphorism νόμος μὲν πάντα κρατύνει, "law governs everything," before launching directly into its theory of generation. The sentence is easy enough to translate, but it is difficult to say much on its philosophical import, standing as it does with no clarifying context or explanation.

29. Watson 1971 situates the natural-law tradition in the Stoics first, and Striker 1987 agrees. Some commentators push the origins back as far as Aristotle (Ilting 1983) or even Socrates (von Leyden 1985). For a discussion of the modern contexts, see Delaney 2003. See also Daston and Vidal 2004; Daston and Stolleis 2006.

30. See Nussbaum 1994; Erler and Schofield 1999.

of the way in which the three classical divisions of Stoicism, its logic, its physics, and its ethics, are so completely interdependent: there was even some debate in antiquity about how to start teaching Stoicism to beginners. Do you start with the physics? But this depends heavily on the logic and even the ethics. Start with logic, then. Now we need to know some physics to get going. Then what about ethics? We still need physics and logic to get it off the ground fully. This is not so much circular as what Dirk Gently would call *fundamentally interconnected* (or strongly coherent, in the modern philosophical jargon). For our current purposes, one of the upshots of this interconnectedness is the very deep relationship between ethics and physics. The Stoics push this interconnectedness in such a way as to ground their entire theory of virtue in their conception of nature. Right morality—and this is the very foundation of what will eventually be called the *natural law* tradition—is simply to live according to nature. As Chrysippus, one of the school's most important authorities, is reported to have said:

> οὐ γὰρ ἔστιν ἄλλως οὐδ' οἰκειότερον ἐπελθεῖν ἐπὶ τὸν τῶν ἀγαθῶν καὶ κακῶν λόγον οὐδ' ἐπὶ τὰς ἀρετὰς οὐδ' ἐπ' εὐδαιμονίαν ἀλλ' <ἢ> ἀπὸ τῆς κοινῆς φύσεως καὶ ἀπὸ τῆς τοῦ κόσμου διοικήσεως.[31]
>
> There is no other or more proper way to approach the theory of right and wrong, or the virtues, or even happiness, than from universal nature and from the administration of the cosmos.

But this intersection of law and nature runs two ways: virtue, happiness, the good, and the bad, are all determined by nature (*physis*). So, too, the cosmos is said to be *administered* or *governed,* and the Greek term, διοίκησις, is a deliberately political analogy.[32]

While there is some debate about whether the Stoics actually invented the idea of a natural law or whether the basic idea predated them, it is very clear that the Stoic theory is much more fully fleshed out than any of its purported predecessors. Nevertheless, we do see that in earlier sources for moral and legal philosophy, ideas about nature had long played important roles.[33] There is, for example, an invocation of something like natural law in Plato's *Gorgias,* where the character Callicles uses νόμος τῆς φύσεως (lit.

31. Plutarch, *St. rep.* 1035C–D.

32. See also Cicero, *De nat. deor.* II.75f. See also Vogt 2008.

33. For example, on the question of whether Aristotle meant *natural law* in the Stoic sense, see Miller 1991.

"law of nature") to justify the rule of the stronger over the weaker. Or from Aristotle:

> λέγω δὲ νόμον τὸν μὲν ἴδιον τὸν δὲ κοινόν, . . . κοινὸν δὲ τὸν κατὰ φύσιν. ἔστι γάρ, ὃ μαντεύονταί τι πάντες, φύσει κοινὸν δίκαιον καὶ ἄδικον.[34]

> I say that law is either particular or universal, . . . [where] universal [law] is that according to nature. For there is a universal justice or injustice by nature, of which everyone divines something.

Whether either of these examples can be seen as a precursor to the full-blown Stoic idea of natural law is still an open question. Nevertheless we can see already in fourth-century BC philosophical sources a bringing together of ideas of law and ideas of nature. This may give us pause: aren't the ideas of law and nature, *nomos* and *physis,* supposed to be fundamentally antithetical in fourth-century Greek philosophy? The answer is no: not necessarily fundamental, not necessarily antithetical, and in any case not universal, as the *Anonymous Iamblichi* and recent readings of Antiphon show.[35]

Human and Divine Governance

There is something in the Aristotle passage that gives me pause. Aristotle's wording is peculiar: the definition of the common and natural law that everyone is supposed to have ready to hand is not "known," it is not "believed," it is *divined.* This is a most curious phrase. Trolling through the

34. Aristotle, *Rhet.* 1373b4–8.

35. Milton 1998 tries at the very outset of his argument to get traction from *nomos-physis* as a "fundamental antithesis" (680), but the whole question of the import, depth, and extent of this so-called antithesis needs to be critically reexamined. The last full-length treatment of *nomos-physis* is Heinimann 1945, and although Guthrie 1962–81 gives the subject 80 pages, his reading is not significantly different from Heinimann's. Variants of this standard version are in Kerferd 1981, chap. 10; and Ostwald 1986; and for a spectacular overinflation, see Kelley 1990. But Gagarin 2002 examines an important set of sources and issues that calls both the sharpness and the universality of the *nomos-physis* antithesis into question (see esp. 65–73, 86–88, 132, 174). Thomas 2000 does an admirable job of contextualizing the set of fifth-century debates about convention versus necessity in sophistic and (especially) ethnographic contexts in which *nomos* and *physis* gain their initial traction as opposed categories (and in these debates it seems clear that *nomos* is primarily understood as *custom* rather than as *law*). But acknowledging that there is such a rhetorical or philosophical context in which some kind of antithesis makes sense need not universalize the dichotomy across time or across discourses. It appears that we would be better off talking of the relationships between *nomos* and *physis,* in many contexts at least, in terms of a continuum rather than in terms of an antithesis.

Thesaurus Linguae Graecae reveals that Aristotle uses the verb μαντεύομαι and its cognates only a handful of times. Discounting the passages where he means it literally, where he is actually talking about divination and portents,[36] we find that in general Aristotle uses the word to mean something very much like *intuit,* where the content of the intuition is meant to be direct and unreflective. But in antiquity, the language of divinations and portents is deeply saturated with a conception of the world in which gods give messages directly or indirectly to humans, and indeed, the gods are never very far away in ancient science. Aristotle himself talks about keeping his physics in line with his "intuition (μαντεία) of divinity."[37] And although I would not want to push Aristotle's uses too hard, it is worth spending a moment on the roles played by ideas of divine governance in ancient theories of law.

In general, I am faced with the problem that, at the hands of later philosophers, the poor Greeks and Romans have never really been left alone with their gods: as early as the third century AD we see the Neoplatonists beating the cream of ancient theology into a lather, the medieval Christian philosophers superimposing a Christian God or disposing of the pagan ones whenever they could manage, and the Enlightenment thinkers, as often as possible, reinventing ancient physics and political philosophy as all-but-godless pursuits. Even today, it is often taken as definitional that ancient science begins where ancient theology ends,[38] and many treatments of ancient political philosophy tend to downplay the foundational roles of the gods, even though natural-law theory is saturated with theology for most of its history.

We saw in the last chapter how Cicero's very influential legal philosophy was grounded in the relationship between the gods and people, but a little more detail will be helpful here. In his *Laws,* Cicero sets out his version of a constitution for an ideal state.[39] The system unfolds in two parallel streams. On the one side, we see him set out the political and judicial organization of legislators, magistrates, judges, and officials of various sorts. But this mundane side gets its justification only through a prior acceptance of a foundational hierarchy that sets priests and diviners up as mediators between the human legislative body and the higher divine one. At the very heart of his *Laws,* Cicero institutes the following hierarchy:

36. E.g., at Aristotle, *HA* 522a17.

37. τῇ μαντείᾳ τῇ περὶ τὸν θεόν, Aristotle, *De cael.* 284b2.

38. See for example, French 1994; Russo 1997; Graham 2006. Contrast Beagon 1992; Flemming 2003; Horstmanshoff 2004; Sedley 2007.

39. On Cicero's ideal system, see also Wood 1988, 70f.

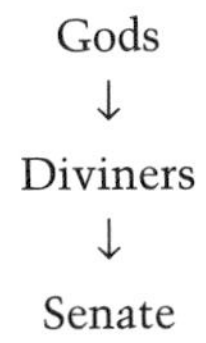

Here the senate, the magistrates, and all the other offices of government are subject to the will of the gods and are enjoined to consult the gods before making important decisions, ranging from declarations of war to domestic and foreign policy decisions, the opening of assemblies, the appointment of magistrates, and so on. The idea is that human government receives its ultimate authority only from the gods, and the gods have veto power over a wide range of human decisions. The diviners thus fill an important role by interpreting what the will of the gods is, and the gods participate directly in the functioning and legitimacy of the state.

Cicero's strategy of situating divination just at the point where human jurisprudence and legislation meet divine jurisprudence and legislation is not the only way of situating divination in a context of law. The important roles played by the *manteis* and by the oracles at Delphi and elsewhere in Greek political history are another example.[40] Earlier still, the Mesopotamians also seem to have seen considerable intersection between human and divine jurisprudence, and this has important consequences for their understanding of the workings of the natural world, and the place of human society in that world.[41] Simply put, human and divine jurisprudence mirror and intersect each other in many ways, and at the divine level the jurisprudential model has important implications for the relationships between humans and the gods, but also for the ways in which the natural world itself is seen as working.

Is a "Law of Nature" Even Possible in Antiquity?

Here a potential problem arises: the idea of natural law may, at first blush, contain in itself a double ambiguity. If we hold to a conception of justice such that *what is just* is *what is according to nature,* then how are we to understand nature without being circular? To see the slipperiness, look at how Gisela Striker puts it in her explanation of Stoic natural-law theory:

40. See, e.g., Bowden 2005.
41. See Larsen 1987; Ritter 2005.

> A good human life, being the life of a rational creature, will have to be organized in accordance with the perfect order of the universe. *Knowledge of the laws of nature* will make one capable of organizing one's life so as to exhibit the orderliness that will make it a good life. Since happiness consists precisely in leading a good life, the Stoics could then even define the good for man as living in agreement with nature.[42]

Notice how Striker slips unwittingly into talk of the laws of nature where the Stoics themselves say only "nature." But there is not necessarily something afoot here: to talk of natural law as coming out of the laws of nature is not to talk in circles, for in this instance what we have are two distinct conceptions of law, if only one of nature. Nature—and what is natural—has to do with the how the universe is constituted.[43] Nature is, for the Stoics, the divine and rational organization of the cosmos as a whole. To say that something is natural is simply to say that it proceeds out of, or is in agreement with, that divine cosmic rational organization. And to say that a *law* is natural for the Stoics means nothing more than that the law derives from, or is in agreement with, the divine rationality underlying the cosmos.[44]

Thus the *law* in natural law refers to the laws that govern human societies.[45] And the *law* in law of nature is, at least in the modern world, used metaphorically,[46] to mean something more like observable regularity in nature,[47] rather than referring to the product of a legislative act (it is just

42. Striker 1987, 91, italics mine.

43. I had originally written "the constitution of nature," but then remembered how ambiguous that may be seen to be in some circles, and notice again the run back into legal terminology. On constitutions, see Latour 1991.

44. See also Daston and Vidal 2004. The old idea that there is a universal taboo against incest plays on just this notion of natural law. But as Inwood has pointed out (1987, 97), natural law is not just about collecting specific universal taboos. It has more to do with how morality gets *grounded* in nature, and so even some apparently universal taboos like incest and cannibalism could be conceived of as acceptable under natural law if the attendant circumstances were just right. On taboo generally, see Douglas 1966.

45. See, e.g., Schofield 1991, 1999.

46. As we shall see, lying behind many historical uses of the phrase *law of nature,* there is the idea of a divine legislator who created the universe and ordained that it follow certain laws, and such cases the use is not metaphorical.

47. Not all philosophers would go in for an unqualified "regularities" reading of "laws of nature," and such language is, to be sure, loaded. But for my purposes at the moment it does not much matter what one's philosophical position *vis à vis* laws of nature is. We will refine the idea shortly. On regularity, pro and con, see Dretske 1977; Armstrong 1978, 1983; Lewis 1983; Cartwright 1983; Earman 1984; van Fraassen 1989; Loewer 1996; Giere 1999; Lange 2000; Chakravartty 2007, chap. 5. Overviews of the major positions are in Carroll 1990; Weinert 1995.

this double meaning of law that Mooney's poster plays on). One now might ask whether law and nature really seem as different from each other as I tried to make them look in the opening sections of this chapter. Perhaps what is most interesting is how these two ideas seem to cover such different realms when we define them, and yet they simultaneously come together so readily. One could argue that the difference lies in precise use versus vague use of the terms, or in literal versus metaphorical, but we would need to be careful what we mean by "vague" or "metaphorical," and whether or when we can impute such qualifications to the historical sources.

Precision is doubly important, since questions around this history of the idea of "laws of nature" within the history of the sciences have tended to focus on the origin of something called *the* modern conception of laws of nature. Although the exact birthdate for this modern idea is still contested, participants in the debate generally agree that it is only since the sixteenth or seventeenth century that we have become fond of describing the regularities in the natural world as operating according to specific mathematical laws. Part of what lies behind this conclusion is a general agreement among historians of science that what we mean by laws of nature are the kinds of things that we find in Newton, but not the kinds of things that we find in, say, Plato, and the debate focuses on exactly where it is between these two thinkers that the idea of descriptive mathematical laws crystallizes. That such a crystallization is almost universally seen as happening more than a millennium and a half later than the sources discussed above is a function of the specific criteria commonly employed for what will count as a law of nature. So also the argument that laws of nature were invented only in the early modern period has been a central part of one standard line of periodization in the history of the sciences. A close look at those arguments will thus allow us not only to situate the legalistic aspects of Roman science with respect to later models, but also to hone our focus on what is implied in calling something a law of nature. I will argue that important criteria for laws of nature are crucially underdefined, and that as we try to refocus the analysis, we find that the early modern origins story misrepresents the Roman material.

Participants in the debate over the origins of the modern concept of laws of nature have often been hesitant to explicitly lay out all of their criteria for what will count as a law of nature in historical sources, and we are left in the awkward position of having to infer the criteria used from the results yielded. The lack of specificity we find in the origins debate is a major problem, since philosophers of science are divided on what an appropriate definition of laws of nature might look like exactly, and on the question of whether there even are or should be such laws.[48] Thus invoking, as the origins debate does, an entity called "the modern concept of the laws of nature" is itself highly contentious. Depending on what *the laws of nature* is taken to mean we run into different challenges to the different positions, and we may even run into challenges to the origins debate overall. As we shall see in what follows, the terms need to be tightened up on a number of fronts for the origins debate to even begin to get off the ground.

Commonly, we find versions of the following two criteria being employed to mark off what will historically count as legitimate "laws of nature":

(a) they should be specific statements that such-and-such *is* a law of nature (as opposed to vaguer statements that there just *are some* laws of nature), and
(b) the *x* in "*x* is a law of nature" should be descriptive, explanatory, and preferably mathematical.

To take an uncontroversial example: on the one hand, Snell's law tells us how light bends when it crosses a boundary between media with different refractive indices. It describes a regularity. On the other hand, Snell's law is a calculation device we can use when trying to determine the precise refraction angle in a medium of known density (or to determine the density of the medium if we know the angles of refraction and incidence). It is a rule we employ in doing optics. So we have two closely related senses of "law of nature." One tells us that nature always behaves in certain ways, and the other offers us a mathematical formula to calculate what we should expect nature to be doing, to solve for unknowns in a natural system. Most of the counterexamples offered from the history of ancient philosophy are dismissed under criterion (a), the *specificity criterion,* since the ancient counterexamples, including the ones we've seen in this chapter, are almost all blanket statements that nature obeys laws, meaning (only?) that nature behaves with regularity. When pushing the specificity criterion we even find that some of the ancient references to the constancy of nature do not actually speak of laws at all, and that law-talk is sometimes an anachronistic gloss on what is actually said in the ancient texts.

Jane Ruby gives us an excellent historical survey of uses of the phrase *law*

48. Dretske 1977; Armstrong 1978, 1983; Lewis 1983; Cartwright 1983; Earman 1984; van Fraassen 1989; Loewer 1996; Giere 1999; Lange 2000; Mumford 2004. For a nuanced historical version of the question. See, e.g., Gordin 2004, 182f.

of nature. As she applies this data set to the origins question, Ruby seems to have three further conceptions of what counts as a modern law of nature that play limiting, if not definitional, roles in her assessment of particular historical uses. It looks as if, in order to count as a precursor of a modern law of nature, a law of nature in a historical source (1) must be purely descriptive, (2) must not rely on an idea of divine legislation, and (3) must "regard nature as a set of intelligible, measurable, predictable regularities."[49] Let us call these three criteria *the Ruby test.* In her analysis, she is able to find all three conditions being met prior to (but not far prior to) AD 1546.

J. R. Milton, on the other hand, wants to push that date forward in time by at least a century. Early on in his argument he objects to Ruby along the following lines:

> One might well doubt whether one of the main explanatory innovations of the scientific revolution [a conception of the laws of nature] was already completed and ready for use even before Copernicus had published *De revolutionibus.* New methodological and explanatory concepts are never fully articulated in advance of their concrete applications.[50]

The idea that a law of nature is "one of the main explanatory innovations of the scientific revolution" is here being used to dismiss Ruby's argument that the innovation preceded 1546, but this just begs the question. After all, wasn't Ruby's argument about whether this innovation even *was* a product of the scientific revolution?[51]

Milton's reasons for dismissing pre-seventeenth-century versions are along the lines that earlier uses of the word *lex* in connection with the natural world were (a) not well defined (alternatively: were "extremely vague"), (b) lacked consensus about what kinds of laws there were, and (c) were not given clearly defined explanatory roles. It seems that what Milton is looking for is a clear statement on the order of "*The law of* refraction is: $n_1 \sin \theta_1 = n_2 \sin \theta_2$," where we have a clearly defined expression of something that is explicitly called a *law.* Let us call this the *Milton test.* The Milton test differs from the Ruby test in emphasizing specificity, explanatory roles, and consensus on what will count as laws.

At a foundational level, Milton's account also seems to rest on some

49. Ruby 1986, 350.

50. Milton 1998, 683. Compare Milton 1981.

51. I am using "scientific revolution" here in an admittedly loose sense, following Milton and Ruby. For hesitations around the term, see Shapin 1996; Park and Daston 2006.

assumptions about the character of ancient science. As he puts it, "There was no room for any such idea [as a law of nature] within the inherited and still intellectually dominant systems of Aristotelian physics and epicyclic astronomy. . . . What was still lacking was a new kind of natural philosophy."[52] If this consideration is normative, that is, if we take the requirement for a new kind of natural philosophy as a definitional prerequisite for the existence of what Milton will count as a law of nature, then the argument is either trivial or question-begging: there were no pre-scientific-revolution laws of nature, because all laws of nature must be post-scientific-revolution. If we take it as descriptive, then the argument depends on us accepting that Aristotelianism is incapable of conceiving of a law of nature, and that epicyclic astronomy is somehow antithetical to laws as well. The reason Milton gives to support the incompatibility of Aristotelianism and laws of nature is this:

> Aristotelian explanations—or rather, explanatory ideals—were essentialist in that they took as their fundamental premises definitions setting out the essences of things. There was no way in which anything analogous to Newtonian laws of motion could be inserted into such explanations, and neither Aristotle nor any of his successors made any attempt to do so.[53]

So there are two preliminary bones of contention: (a) Aristotle offers "explanatory ideals" rather than explanations, and (b) those ideals take definitions as their fundamental premises. But I'm not sure that either of these points would be granted as they stand by anyone who has read Aristotle carefully, and in any case, the second point is certainly not any support for Milton's argument, since Newton himself takes a set of definitions as his fundamental premises in the *Principia*. Milton's third and most central objection, that of essentialism, is a common enough characterization of Aristotle, and even though such a claim would be seen by many Aristotle scholars as subject to at least some qualification, it must be on this point that Milton's argument is seen to hinge. Thus if Milton's claim is to hold, it needs to be shown that the following pair of propositions are mutually exclusive:

(1) Things have essences.
(2) Natural regularities can be described by laws.

52. Milton 1998, 684.
53. Milton 1998, 680.

Exactly how to unpack (1) in terms of Aristotelian philosophy has been the subject of considerable debate,[54] but I fail to see how any of the proposed readings of Aristotelian essentialism conflicts with (2), and if Milton's argument is to hold, the case that one or all of them does so needs to be made explicitly.

I am not here claiming that Aristotle *did* have a conception of laws of nature that corresponded in any significant way with Newton's, but only that there is nothing about Aristotle's philosophy that has been shown to preclude such laws. Milton himself is being an essentialist of a sort, and the incompatibility he wants to see between Aristotelian philosophy and laws of nature remains crucially undemonstrated. What is worse, it may even be the case that a degree of essentialism is desirable or even necessary on some modern understandings of the meaning of laws of nature.[55]

What about the other "intellectually dominant system" before the scientific revolution, that of epicyclic astronomy? Without telling us what he means by the *system* of epicyclic astronomy (and to lump Ptolemy, al-Tusi, Copernicus, and Tycho together under the rubric of a single "system" is contentious in the extreme), Milton offers the following explanation of the incompatibility of that system with a conception of a law of nature:

> Newton's laws are part of the central explanatory core of the *Principia,* and the equivalent of this in [Copernicus'] *De revolutionibus* is the Hipparchan geometrical apparatus of epicycles and eccentrics. *De revolutionibus* could very easily be rewritten so as to exclude any mention of laws, and the basic content of Copernicus's theory would be quite unchanged.[56]

This needs unpacking. In every edition of the *Principia,* Newton begins with two important sections: (a) *Definitiones,* and (b) *Axiomata, sive leges motus.* The definitions are numbered, *Def.* I, *Def.* II, and so forth, and the Axioms or laws are numbered *Lex.* I, *Lex.* II, and so forth. Beginning a work with definitions and axioms is, though, nothing new at all. It is at least as

54. See, for example Bennett 1969; Marcus 1971; Brody 1973; S. M. Cohen 1978; Lloyd 1987; Matthews 1990; Ferejohn 1991; McKirahan 1992; Barnes 1994; Wedin 2000; Tierney 2004.

55. See Carroll 1990, 196–97. As Anjan Chakravartty has put it to me: "If things have kind essences (modern translation: sets of properties, each of which is necessary and which together are jointly sufficient for kind membership), then certain generalizations about the natures (e.g. characteristic constitutions) and behaviors of such things are *guaranteed* to be lawlike, since such things have some properties necessarily."

56. Milton 1998, 684.

old as Euclid, and it is common in most branches of ancient, medieval, and early modern mathematics, including what we might call applied mathematics: optics, statics, mechanics. So what is significantly new in the *Principia*? There are three possible candidates: (1) Newton's equivalence of *axioms* with *laws*: *axiomata*, SIVE *leges motus*. (2) The specificity of the phrase laws *of motion*. (3) The content of the laws themselves.

If (1) and (2) are conceived as not being reducible to (3), then the question of their significance is strictly speaking either lexicographical (as in case [1], where we see a definitional equivalence) or philological (as in case [2], where we see the use of *lex* plus the genitive of a particular class of noun), and it is lexicographical or philological only. But if the point is that the terminology or the grammar is supposed to point to something deeper, to a fundamental innovation in the conceptualization of nature, then the shift in terminology or grammar has meaning only in terms of (3), the content of the laws themselves. This point should not be taken lightly. Both Ruby and Milton share the assumption that what we should be looking for in prescientific-revolution sources as precursors for law-of-nature accounts, will be marked by explicit uses of the phrase *law of x*. Now if on the one hand, the ancients have something that fits all the conceptual definitions of what we call a law of nature but don't call it a *law*, and if on the other hand, when they do use the phrase *laws of nature* they do not mean what we call *laws of nature* under the Ruby and Milton tests, then the question of the origins of "scientific law" is one of a shift in definitional terms, not one of a shift in conceptual frames.[57]

But if we move instead to a content-based analysis, then we open up the possibility that in the Middle Ages or antiquity there *were* statements about the behavior of the natural world whose content and form were similar to or identical with the early modern *laws of nature*, even if the medievals or the ancients did not use their words for *law* to describe those statements.

Taking a content-based approach, we will need to show whether, why, or how Newton's laws are qualitatively different from, say, Ptolemy's explanation of planetary motion in terms of the equant or even his statement of uniform circular motion in the first place. To be fair, Ptolemy's argument in general proceeds planet by planet, but we do have many examples where he invokes similar rules for each planet individually—or even better, where

57. Ruby's dismissal, for example, of the possibility of a Ruby-test-passing candidate from medieval Islamic science strikes me as trite: "Arabists," she says, "tell me there is no word corresponding to "law" in Arabic" (Ruby 1986, 344n19). If the history of the *Frisbee* cannot be pushed any farther back than the changing of its name from *Pluto Platter*, then we are talking about the history of a word, not of a thing.

he sets down rules that explicitly apply for every planet, and I am not certain that these are different in kind from the laws Newton offers us. They are worded differently, to be sure, but part of that is a function of changes in the conventions of mathematical writing between Ptolemy and Newton. Ptolemy may be verbose, but he is neither vague nor unmathematical. Take the following, which we might call the *Ptolemy's ᵀlaw of planetary station,* where by *ᵀlaw,* I mean "tentative law":

> ἐάν τε διὰ τῆς κατ' ἐπίκυκλον ὑποθέσεως γίνηται τοῦ μὲν ἐπικύκλου περὶ τὸν ὁμόκεντρον τῷ ζῳδιακῷ κύκλον τὴν κατὰ μῆκος πάροδον εἰς τὰ ἑπόμενα τῶν ζῳδίων ποιουμένου, τοῦ δὲ ἀστέρος ἐπὶ τοῦ ἐπικύκλου περὶ τὸ κέντρον αὐτοῦ τὴν τῆς ἀνωμαλίας ὡς ἐπὶ τὰ ἑπόμενα τῆς ἀπογείου περιφερείας, καὶ διαχθῇ τις ἀπὸ τῆς ὄψεως ἡμῶν εὐθεῖα τέμνουσα τὸν ἐπίκυκλον οὕτως, ὥστε τοῦ ἀπολαμβανομένου αὐτῆς ἐν τῷ ἐπικύκλῳ τμήματος τὴν ἡμίσειαν πρὸς τὴν ἀπὸ τῆς ὄψεως ἡμῶν μέχρι τῆς κατὰ τὸ περίγειον τοῦ ἐπικύκλου τομῆς λόγον ἔχειν, ὃν τὸ τάχος τοῦ ἐπικύκλου πρὸς τὸ τάχος τοῦ ἀστέρος, τὸ γινόμενον σημεῖον ὑπὸ τῆς οὕτως διαχθείσης εὐθείας πρὸς τῇ περιγείῳ περιφερείᾳ τοῦ ἐπικύκλου διορίζει τάς τε ὑπολείψεις καὶ τὰς προηγήσεις, ὥστε κατ' αὐτοῦ γινόμενον τὸν ἀστέρα φαντασίαν ποιεῖσθαι στηριγμοῦ.
>
> If [the synodic anomaly] is represented by the epicyclic hypothesis, in which the epicycle performs the [mean] motion in longitude on the circle concentric with the ecliptic toward the rear [i.e., in the order] of the signs, and the planet performs the motion in anomaly on the epicycle [uniformly] with respect to its centre, toward the rear along the arc near the apogee, and if a line is drawn from our point of view intersecting the epicycle in such a way that the ratio of half that segment of the line intercepted within the epicycle to that segment intercepted between the observer and the point where the line intersects the epicycle nearer its perigee is equal to the ratio of the speed of the epicycle to the speed of the planet, then the point on the arc of the epicycle nearer the perigee determined by the line so drawn is the boundary between forward motion and retrogradation, so that when the planet reaches that point it creates the appearance of station.[58]

Unwieldy as it is, we could, if we like, rework this law into a slightly more modern notation, as in figure 3.2. Now, does this pass the Ruby test? Yes: it is descriptive, it does not invoke divinities, and it is certainly treating

58. Ptolemy, *Alm.* XII.1, 450–51. Translation very slightly modified from Toomer 1984, 555. The sentence of which this is a clause is truly stunning: it runs to 36 lines in the Greek.

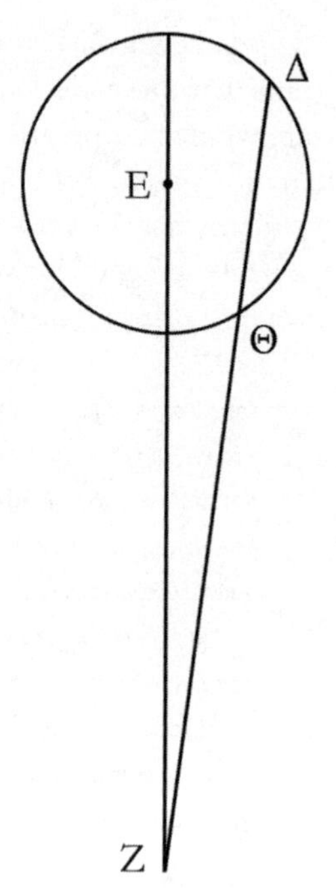

Fig. 3.2 For any planet, given a line ZΘΔ drawn from the earth (Z) to the epicycle, and intersecting the epicycle at points Θ and Δ, and given the ratio of the speed of the epicycle to the speed of the planet, A : B, then the planet will appear stationary at Θ when

$$\tfrac{1}{2}\,\Theta\Delta : Z\Theta = A : B.$$

nature's regularities as intelligible, measurable, and predictable. So on all three criteria, Ptolemy's ^{T}law is a ^{Ruby}law. But is it a ^{Milton}law? Certainly the ^{T}law is well defined and nonvague. I am not entirely certain whether it has a "clearly defined explanatory role" on Milton's terms (his criteria are underdefined), but I will say that it strikes me as explanatory in the same way as many of Newton's laws of planetary motion. As for Milton's consensus clause, it is clear that both Ptolemy's ^{T}law and its proof are very much in line with the kinds of things going on in earlier and contemporary mathematical astronomy.

One possible line of rejoinder I can see here would be an objection along the lines that Newton's laws (sometimes) invoke physical forces in a way that Ptolemy's don't. Certainly Newton seems at first to talk this way:

> Gyretur corpus in circumferentia circuli, requiritur lex vis centripetae tendentis ad punctum quodcunque datum.[59]

> Let a body revolve in the circumference of a circle; it is required to find the law of the centripetal force tending toward any given point.

59. Newton, *Principia* I.2.7.2.

If we look at what Newton actually shows in this proposition, we see that the *law of the force* that he finds is, in the end, mathematical: the law is the relationship between the distance between the current position of a circularly moving body and the position it would have occupied had it moved in a straight line to a line drawn in a specific way from any random other point. The *law* of force Newton is talking about here is a mathematical relationship between lines, describing the trajectory of a moving body.

Nonetheless, Newton does call this mathematical relationship a law of *force,* and this is something we would be hard pressed, I think, to find Ptolemy doing, even if we interpret "force" analogously. On a modern understanding, we see the word *force* as referring to some physical chain of events operating between the bodies and bringing them into interactive relation. This chain of events can be described mathematically using Newton's law of centripetal force, but the idea of a force underlies the mathematics in some way, such that the mathematics is seen as describing and quantizing that force. But neither Milton nor Ruby has insisted that real laws of nature must have a force interpretation immediately underlying them, and they would be ill advised to do so, since many of what they would call laws of nature cannot be interpreted in terms of nonmediate forces. Kepler's area law is one, Snell's law another, entropy a third. Pressing the force interpretation here is only to point out that Newton is doing physics in this proposition, and Ptolemy is not. In astronomical propositions, we see Newton invoking laws that are no different in kind from Ptolemy's ^{T}law of planetary station, or the rules around how equants work, or how mean planets move in relation to the mean sun in *Almagest* X.6, to pick a few examples.

The second potential attack on the lawlikeness of Ptolemy's ^{T}law runs along the lines that it tells us only about the appearances of particularly situated bodies. It is a truth about geometry, not about the natural world. If this is to fly as a crucial distinction between Ptolemy's ^{T}law and Newton's laws, then we need to find some way in which Newton's laws demonstrate a truth-about-the-natural-world quality that supervenes on geometry in a way that Ptolemy's ^{T}law doesn't. Let us try a test case: if we take Kepler's area law, for example, what is more "natural-worldy" about it than what we find in Ptolemy? As Newton handles it, the area law states that a line drawn from a body moving in an elliptical orbit around the sun as one focus of the ellipse will, if it is being impelled toward the sun by an inverse-square force, trace out equal areas in equal times. For his part, Kepler had left out the inverse-square attraction to the sun, and simply posited the area law in order to define the relative speed of the planet at any point on its orbit. If

Kepler's version of the area law is taken as a true law of nature, then so must Ptolemy's definition of planetary motion in terms of equants. The two are exactly symmetrical, and Ptolemy's equant offers us a ^{Kepler}law. If we deny Kepler true lawhood and give it instead only to Newton for his very clever integration of inverse-square attraction, then what are we doing exactly? We are saying that the interaction of two distinct motions, *viz.* rectilinear centrifugal and inverse-square centripetal, will generate a specific geometrical relationship, *viz.* equal areas. But this is not substantively different from Ptolemy's ^{T}law, which states that the interactions of two distinct motions, *viz.* circular epicyclic and circular deferential, will generate a specific geometrical relationship, *viz.* apparent station at a given point. What Ptolemy's ^{T}law of planetary station gives us is an elegant equation that holds between two pairs of relationships: epicyclic and deferential speeds on the one hand, and epicyclic and deferential sizes on the other. So, too, Newton gives us a relationship between planetary speed and planetary distances. Is the appearance of station any less *real* than the line that sweeps out equal areas? Is circular motion about epicycles and deferents supposed to be less *physical* in Ptolemy's conception than the rectilinear motions are in Newton's? Evidence from Ptolemy's *Almagest,* where he appeals to Aristotelian physics, would suggest otherwise, as would the physical realism that, as Bernie Goldstein showed, underpins Ptolemy's *Planetary Hypotheses.*[60] Finally, any appeal that tries to call into question the relevance of apparent planetary stations to laws of nature anachronistically downplays the relevance of apparent stations as natural phenomena in antiquity.[61] And are the laws that relate temperature and humidity to windchill factors any less about appearances? What about the relationships between apparent size and actual distance in optics?

Divinity, *Redux*

In looking for the origins of the modern concept, Ruby rules out any account that has recourse to a divine legislator. I can see no valid reason for her doing so, and Milton has convincingly shown both that there is no contradiction between deism and physical laws, and that many of the major scientific revolution figures who are credited by both him and Ruby with full-blown law-of-nature accounts did, in fact, rely on divine legislation at

60. Goldstein 1967; See also Musgrave 1981; Jones 2005; Evans and Berggren 2006.

61. Indeed, they are one of the natural phenomena now thought to have been tracked by the remarkable Antikythera mechanism. See Freeth et al. 2006; Evans, Carman, and Thorndike 2010.

one level or another.[62] This has the effect of highlighting an imprecision in some modern readings of historical law-of-nature accounts that see the *law* in law of nature as metaphorical,[63] but if there is a divine legislator, then the understanding of law is literal, not metaphorical. Keeping this in mind will allow us to see the breadth of what was meant by *law,* rather than imposing modern constrictions on premodern terms.

Milton's point about the interrelationships between divinity and laws of nature is in keeping with recent approaches in the history of the sciences that recognize the important roles of theology in the history of science generally. Right up into the modern period, there is often an intricate interplay between theology and the sciences. Newton's theology and eschatology have recently received a fair bit of air in both professional journals and in the mass media.[64] Descartes' universe can't even get off the ground without the guarantees provided by a nondeceptive and unchanging God. When push comes to shove, all of Aristotle's physics relies on the primary motions provided by the prime mover who is (crucially) θεῖον, divine. Pliny begins his *Natural History* with an ode to the deity that provides order to the universe, and Aratus' *Phaenomena* opens with the line "Let us begin with Zeus . . ." Stoic physics is saturated with divine rationality. Even the Epicureans, who argue that the gods do not interfere in the running of the cosmos in any way at all, emphasize that the gods exist. They do not, indeed cannot, play any roles in the workings of the cosmos (it would be beneath their dignity and perfection to be meddling with changeable matter), but they play important roles in Epicurean ethics, where they stand as moral models for human happiness. The gods then give us an indirect route to physics, insofar as the ethical happiness they instantiate serves as the basic drive to the elaboration of physics in the first place: the elimination of fear. To be sure, this—and it is emphatically the extreme case—puts the gods at one remove from physics, but they are never completely detached.

But to say the gods often have important roles in the historical sciences is not yet to parse out precisely what those roles are, and it is here that much work still needs to be done. If gods create the universe and its laws, that is one thing; if they enforce those laws on a day-to-day basis, that is quite another. And in Descartes, for example, we see God playing a different role again.[65] For him God acts as a kind of epistemological stopgap for ques-

62. Milton 1998, 695f.

63. E.g., Ruby 1986 calls it a metaphor throughout her paper.

64. On Newton, see, e.g., Iliffe 2007; Snobelen 2001.

65. Descartes 1664.

tions about the constancy of laws of nature. In *Le monde,* Descartes grounds our knowledge of the unchangeability of the laws of nature in the eternal unchangingness of God: since the laws were created by God, and God is constant and eternal, then we know that the laws at one time or in one place are the same always and everywhere.

But if the gods have important roles in the natural order—whether their roles be foundational, epistemological, participatory, or all three—then any account of the natural order has to include the gods. The implications here for the current debates are that, again, Ruby's *no-gods* criterion is untenable and anachronistic.

Conclusion

I have argued that questions about origins need to use a content-based analysis. Rather than simply looking for statements of the form "*x* is a law of nature," we should look for lawlike statements that nature does *x*. In order to do this, we will have to be very clear about what our criteria for *lawlikeness* are, without failing to pay close attention to the problems inherent in different available definitions.

It is true that we see some uses of the phrase *law of nature* in antiquity and the Middle Ages, but Ruby and Milton are right that many of these uses will not fit most, if not all, modern definitions of what a law of nature is. Nevertheless the phrase *law of nature* itself does not ring hollow in antiquity—far from it. If we look for ancient sources that juxtapose conceptions of law and conceptions of nature, we find that those juxtapositions are neither rare nor superficial. In order to highlight some of the problems facing us, I floated the possibility early in this chapter that laws and nature could be conceived in such a way as to make their conflation look unusual or strained. But my making the two concepts look foreign to each other turned out to be artificial when our look at the historical record found two quite common ways in which the concepts of law and nature come together. Law and nature are not so distant from each other after all. The upshot of this is that it should occasion no surprise that the kind of mathematized and universal regularity pointed to in Ptolemy and other premodern sources should come eventually to be called *laws* of nature. Indeed, the differences between Ptolemy's law and what Lucretius calls "laws of nature" probably have more to do with the depth and kind of investigation each author was engaged in than in any fundamental disagreement over whether nature was, as we would say, lawlike. There is nothing novel in calling natural regularities *laws,* and in any case, what really matters is not the terminology, so much as what the terminology might imply. Again, this points back to the importance of a conceptual analysis rather than a lexicographical or philological one.

In many of the ancient cases we have surveyed, conceptual frames that run ideas of law and ideas of nature in parallel are ultimately rooted in worldviews that see gods as foundational in some way, and none of our authors—not even the Epicureans—deny them outright. In many accounts, we see gods legislating and/or guaranteeing the constancy and universality of natural processes. But we should note that divine legislators are not a necessary prerequisite for talk of nature as lawlike in antiquity. Lucretius' nature is constrained by what he calls laws, but it is at the same time completely free from divine intervention. The other front for the coming-together of law and nature is in the natural-law tradition where, again, divine governance is a common foundation for talk of how human governance should work. But, just as the gods can have quite different roles in different accounts of nature as lawlike, so, too, do they play different roles in different natural-law theories.

I would like to offer a few qualifications to what I have argued in this chapter. I showed that if we drop the emphasis on terminology and instead focus on content, then the ancient sciences may well offer us many lawlike accounts of nature. The detailed examples I used were for the most part from physics and astronomy, but this choice was largely a function of the particular debates I was engaging. We should also keep in mind how differently ideas of lawlikeness play out in fields outside the physical sciences. In general, the standard emphasis on physics and astronomy in the origins debate may be more of a problem than is at first obvious. Philosophers of the life sciences will be all too familiar with the tendency, until very recently, for modern accounts to treat physics as *the* paradigmatic science. Rich and complex philosophies of science have been developed that turn out not to be easily adapted to biology without assuming a harsh reductionism. One is reminded here of the old barb (attributed to the physicist Ernest Rutherford) that "in science there is physics; everything else is stamp collecting." Philosophers of biology have been reacting against this conception for some time now, and recent work has begun to focus on possible problems with expecting biology to obey the kinds of specific laws that physics is so fond of.[66]

66. See, e.g., Beatty 1997; Brandon 1997; Sober 1997; Waters 1998; Mitchell 2000; Woodward 2001; Creager, Lunbeck, and Wise 2007. For a different view on biology and laws, see, e.g. Lange 2000.

But this point runs further. We also need to include in the discussion historical practices that have no modern analogues, like astrology and divination.[67] We shall see throughout the present book how rich the arguments, epistemologies, and ontologies turn out to be for many of the ancient ideas that no longer have currency in the canon. Moreover, what we might call the legalization of nature can happen in several different dimensions, and the question so far has looked only at two. As we shall see in the subsequent chapters, legal models can also serve to formulate and underpin (and also to sometimes undermine) the degrees of certainty that we can hope to attain in our understanding of the natural world. *Evidence*, after all, has consonant meanings.

67. E.g., any origin arguments will need to take a good hard look at the paradigmatic Babylonian science, extispicy (entrail divination): are the general statements in the first tablet of *Multābiltu* to be taken as lawlike or not? (See Koch-Westenholz 2000, 23; Starr 1983, 9f.) Here anachronism becomes a real danger, for both the *pro* and *con* sides. But if we attend to contemporary cultural, cosmological, and epistemological contexts, we will see that it becomes difficult to maintain that they are not lawlike without adopting apparently arbitrary constraints that privilege modern criteria as standards. This is a problem that Carroll 1990, 201–2, and Roberts 1999 have pointed to in Lewis' account of laws (see Lewis 1983).

CHAPTER FOUR

Epistemology and Judicial Rhetoric

We have seen how nature was taken in natural-law theories to underscore ethics, law, and good governance, and at the same time how ideas of lawlikeness pervade conceptions of nature. This dual grounding of the law/nature continuum gets a fuller fleshing out as we follow it into new territory in the early principate, where law not only provides models for how regularities in nature work, but where legal theory also gets used to both ground and qualify the certainty attainable in investigations into nature. Here legal practice becomes a new kind of epistemological tool. As a correlate of this, we will also see how observation of the natural world—one of the very foundations of the sciences—comes to be handled as an ethical problem.

In the present chapter I explore how, in the century or so after Cicero's death, rhetorical structures and a rhetorically-based judicial epistemology come into play in natural investigations. Where Cicero had been frequently pushing a very strong conservative political agenda in the last days of the republic (mid-first century BC), such an agenda ceased to be viable as the empire settled into a more or less stable principate under the early emperors. Much has been written on the political aspects of literature under Augustus, with a general agreement that the kinds and degrees of political participation possible under his regime and those that followed meant that political activities among the elite—and the literature and performances that enabled or carried those activities forward—took on new and often very subtle forms.[1] Even fairly early in his career, the philosopher and dramatist

1. See, e.g., Woodman and West 1984; Habinek 1998; Roller 2001; Volk and Williams 2006; Davis 2006; Connolly 2007, chap. 6.

Seneca had found himself at the pointy end of imperial politics and knew firsthand how dangerous the game could be. He had, according to one story, been condemned to death by Caligula in AD 41 for a speech he made in the senate (Dio hints that the problem was not the content of the speech but its quality: the emperor was jealous of Seneca's brilliance),[2] and he had gotten off only because Caligula was told (falsely) that Seneca was dying of consumption anyway, and so there would be little point in expediting his demise. Moreover, Seneca incurred the wrath of more than one emperor, being banished on a (politically motivated?) charge of adultery under Claudius. Nevertheless, he found his way back to the center of things, first as tutor, and then as chief advisor to a young Nero, who took the reins of the empire in AD 54. By the time Seneca had come to write the *Natural Questions* in the early 60s, though, he had been forced out of politics as Nero had become increasingly erratic and the balance of power at court had shifted away from the moralizing Seneca. Seneca's philosophical focus during his retirement was largely concentrated on ethics very broadly rather than governance and power in particular. Retirement or no retirement, though, Seneca ended his life at Nero's command—political intrigue has long hands and a long memory.

Looking to Seneca's works on nature, we find ethics front and center. This makes sense only if we look to the overall cultural, ontological, and taxonomic structures within which Seneca's investigative project operates, beginning with a look at his use of vocabularies of comparison and example.[3] The idea that develops from this is that the selection of which observations will serve as examples—and building on this, the crystallization of particular scientific objects as *things* to be studied, discussed, observed, and argued about—can be fully understood only if we look at both formal and epistemological features of the discourse in which scientific examples and objects are embedded.

As we dig into Seneca's strategies for introducing entities and facts into his discourse on nature, we will see not only how thoroughly moral his concerns with nature are, but also the extent to which morality enters into the acceptability of observation claims. The eyes, for Seneca, are deeply ethicized. Not only is seeing conceived as a moral act (as it had been already in Lucretius), but (following standard courtroom practice) moral character is taken explicitly to inform the authority of individual witnesses.

2. Cassius Dio, LIX.19.

3. On the uses of example in Greek philosophy, see Lloyd 2004, chap. 9.

Theory-Ladenness and Observation

Questions about the relationships between observations and theories have long been a subject of careful and sometimes passionate argument in the philosophy of science, as well as in perceptual psychology. In philosophy, the sharp distinctions initially drawn by logical empiricists came to be questioned in the middle part of the twentieth century on the grounds that observations seemed to be necessarily theory-laden, and so not sharply distinguishable from theory after all.[4] This seemed to many to imply that observations may not provide objective support for the theories with which they are already laden, and extreme relativism looked to be lurking in those shadows. In the 1980s Jerry Fodor tried to salvage objectivity for observation by arguing that perceptual systems (vision, smell, touch) were *modular* and so were insulated from certain kinds of theoretical intrusion—we may know that the two lines in the Müller-Lyre illusion are the same length (see fig. 4.1), but that does not help us to see them as the same length, and therefore our vision is nonmodifiable by theory at some level.[5] On the other hand, recent work in cognitive psychology has argued that very few perceptions are of this sort, and that preconceptions *can* be shown to measurably influence perception of a wide range of phenomena.[6] It's not that Fodor was wrong about optical illusions, but that optical illusions are atypical perceptions.

It is no accident that many of the philosophical discussions of theory-ladenness draw on historical evidence. In historical sources, theory-ladenness stands out with particular starkness. There are several reasons for this. One is that the very different theoretical commitments that we find in historical sources—from final causes to N-rays to phlogiston—make observational selectivity, approximation, and error all the more obvious and (it would seem) easily attributable. The second is the nature of the sources: the farther back we go, the less likely we are to have any bare observation reports, lab notes, or rough work. Instead we have to rely more and more on final versions of observations as always already synthesized into arguments.[7] We should

4. Classic loci, pro and con, include Duhem 1906; Hanson 1958; Kuhn 1962; Putnam 1962; Hempel 1965; Carnap 1966; Feyerabend 1975; van Fraassen 1980; Churchland 1979, 1988; Fodor 1984, 1988.

5. Fodor 1983.

6. A very cogent summary of this work, including a nicely balanced treatment of the Fodor example, can be found in Brewer and Lambert 2001.

7. The most obvious ancient exception to this claim proves most interesting: what are we to make of all the "bare" observations in the Hippocratic *Epidemics I*? How theory-free is the observation that a fever spiked on day 17? See Baader and Winau 1989; Langholf 1990.

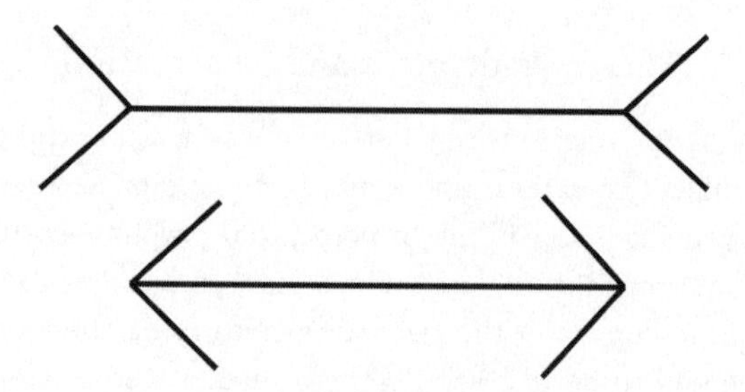

Fig. 4.1 The Müller-Lyre illusion

worry that this means that the most basic level of question about the degree to which observations are theory-laden *at the moment of observation* in these sources may be inaccessible or else question-begging. But this limit is not terminal, as we shall see.

A useful starting point is to look at the ways in which theories select for observations, and the ways in which observations get employed to convince readers of a larger point. To do this, I want to follow an extended argument and watch for when and how the author uses observation claims. Theory-based selectivity plays a large, but by no means the only, role in these choices. It is not only the theoretical entanglements of individual facts that matter, but also the contexts of the arguments themselves: the political, cultural, and educational contexts of ancient scientific inquiry leave an indelible stamp on the modes, contents, and limits of that inquiry. Rhetoric in particular, the cornerstone of Roman upper-class education and one of the single most important skills in the Roman intellectual tool kit, plays some especially interesting—and rather surprisingly large—roles.[8] As always, the danger that we import our own categories onto historical investigation is very real, and we need to resist the urge to impose modern limits on what counts as rhetoric. Rhetoric does not just cover modes of arguing in antiquity, but also houses many of the ancient theoretical discussions of the logic of scientific discovery, to borrow a handy phrase. Many historical discussions of the epistemology of the sciences, of scientific method, of induction, of empiricism, have unfortunately limited themselves to what can be found in ancient philosophical and logical texts. But it is in the rhetorical texts that many of these problems get their fullest fleshing-out, and the

8. On the changing roles of rhetoric in Roman education between the late republic and early principate, see Bonner 1977; Morgan 1998; Kaster 2001; Corbeill 2001; Too 2001; Connolly 2007; Compare Bloomer 1997.

ignoring of the rhetorical discussions has unfortunately skewed our picture of the historical development of the sciences.

To point out that an ancient scientist was engaged with rhetoric is not novel.[9] Geoffrey Lloyd influentially argued a more general case for a rhetorical and jurisprudential core to the very contexts of Presocratic philosophy and Hippocratic medicine.[10] But Lloyd also pointed to a real tension at the heart of this relationship in Greece, where philosophers were long uncomfortable with their relationship to rhetoric, and much effort was expended on trying to distinguish philosophy from "sophistry," in spite of the very strong overlaps that even contemporaries were aware of—as is apparent in Aristophanes' treatment of Socrates in the *Clouds*, for example.[11] In the Roman imperial context, though, rhetoric has become so thoroughly central to the wider intellectual culture that the problem of its boundaries vis à vis philosophy becomes significantly less of an issue. Anthony Long credits the shift to Cicero:

> Cicero recognizes the difference between philosophy and rhetoric, but he constantly insists on the desirability of combining them, to the mutual advantage of each. That theme, which is one of his distinctive contributions, strongly influences his manner of writing philosophy. What applies to Cicero is also pertinent to Roman philosophy more generally.[12]

An excellent case in this regard is Seneca's oft-overlooked *Natural Questions*. One gets the impression that the time is passing in which one needs to offer an apologetic before opening an extended inquiry into the *Natural Questions*. In the wake of work by Brad Inwood, Harry Hine, Gareth Williams, and others the text is not only more accessible, but also considerably richer than it had long looked to be. Very widely read right up into the early modern period and of considerable importance to the history of the sciences, it eventually fell into obscurity. I might also add that the *Natural Questions* is the only extended treatise on physics still extant from the pen of a professed Stoic. It has often been remarked that Seneca wears his Sto-

9. A strong case has been made for Pappus of Alexandria in particular. See Bernard 2003. See also Cicero, *Top.* 81–82, where he cleaves off a genus of rhetorical inquiry called *quaestiones infinitae cognitionis*. On Cicero's innovations in rhetoricizing philosophy, see Douglas 1995; Long 2003. Contrast Dominik 1997, 53, which offers a more limited reading of Quintilian, XII.2.

10. Lloyd 1979, chap. 2.

11. See esp. Lloyd 1990, chap. 3.

12. Long 2003, 192.

icism lightly and is not concerned to cleave to or develop all the minutiae of school doctrine. Indeed, he is often very critical of it. Seneca is not, in this respect, an ideal source for an ideal of Stoicism. That doesn't matter so much for me. He is exceptionally interesting as an individual thinker, and we *can* take him as a representative of a different kind of Stoicism, that is, of philosophy in the flesh, as it walked and talked in the first century, in Nero's Rome. This is a Stoicism immersed in the petty details of politics, of day-to-day life, of weird Roman interests in volcanoes and caverns, comets and rivers, and, above all, a Stoicism that is not aspiring to ideological purity but instead feels free to borrow from other schools and to find novel solutions to old problems.

The *Naturales quaestiones* has been translated into modern languages several, but not all that many, times. The title does some interesting things. The English and French translations opt for homonymy: *Natural Questions* and *Questions naturelles.* The recent Italian translation renders it as *Ricerche sulla natura,* and the German as *Naturwissenschaftliche Untersuchungen.* So what is a *quaestio?* A question, an investigation, an inquiry, to be sure; but these broader senses are underscored in Latin (as in the modern languages) by a more context-specific meaning of the word: a judicial inquiry. Seneca was never one to avoid playing with rhetorical forms in any genre, prose or poetry,[13] so we would do well to keep the idea of inquiry, rather than the more mundane *question,* at the fore. As we shall see, Seneca gives us abundant cues in the *Naturales quaestiones* that he frequently has not just a rhetorical model, but in particular a *judicial* model, in mind. These cues range from the recurring use of explicitly jurisprudential language, to his handling of witnesses and evidence, the conscious use of what are called the *figures* of judicial rhetoric, and the more subtle but still clear imposition of structure on the overall arguments of each book of the work. It is in many ways as though the phenomena of nature itself are on trial, and Seneca has published his speeches to the court. This reading also sheds some much-needed light on Seneca's shadow interlocutor, whose voice plays such an important but sometimes puzzling role in the *Natural Questions.* Here Seneca's novelty lies not only in the extent to which he brings specifically judicial arguments to bear on his subjects, but especially in the epistemological consequences of this bearing: in offering his argument as though it were before judges, he is consciously pushing the best case he can muster, rather than seeking to pave the world with finished doctrine.

13. See, e.g., Kennedy 1972, 469f.; Goldberg 1997; Wilson 2007.

Observations as Models

Inwood has argued, and I think he is right, that at the heart of Seneca's *Natural Questions* is an inquiry into the relationships between god and man.[14] The higher division of this study, that pertaining to the gods, is not limited to what can be seen: *non fuit oculis contenta.*[15] If this is a play on Lucretius' earlier *nulla potest oculorum acies contenta tueri* (I.324), it is a clever one, insofar as *contentus* has been dragged across from one verb (*contendo*) to another (*contineo*). Nonetheless, the semantic stretch happily still works to make the same point: the really important stuff, what is fundamentally true, is accessible only to reason, not to the eyes: *nihil esse acie nostra fallacius*, "nothing is more deceptive than our eyesight";[16] *visus noster solita imbecillitate deceptus*, "our vision, deceived by its usual weakness,"[17] and so on. It is notable, though, that Seneca (like Lucretius before him) does in fact spend a good deal of time discussing things that can be seen with the eyes, and the eyes form a most important tribunal. As Lucretius says, *quid nobis certius ipsis sensibus esse potest, qui vera ac falsa notemus?* "What could be more certain than our own senses, to distinguish truth from falsehood?"[18] And, for all that the real truths are hidden (think again of Cicero's higher-order observations), they are still experiential in some way, as Seneca's definition of the cosmos itself shows: it is "everything that falls or can fall before our notice": *omnia quae in notitiam nostram cadunt aut cadere possunt mundus complectitur.*[19] So both the *Naturales quaestiones* and the *De rerum natura* are rich in observational examples, even if the ultimate truths transcend the eyes in both.[20] This tension is not irresolvable, as we shall see, but it does put pressure on methodologies and on the rhetoric of science.

So what do observation statements do in physics? We see in the first instance that observation statements provide mundane illustrations or models for how supermundane phenomena happen.[21] They thus simultaneously

14. Inwood 2005, chap. 6. Inwood uses "relationship" in the singular, but I want to flag the plurality here.

15. *NQ* I.praef.1.

16. *NQ* I.3.9.

17. *NQ* I.2.3.

18. Lucretius, I.699–700. Compare IV.478f.

19. *NQ* II.3.1.

20. On a similar tension in Greek philosophy, see Lloyd 2004, chap. 9; Ierodiakonou 2002.

21. Comparison to the mundane and familiar also, to use Williams' word, *domesticates* otherwise frightening phenomena to easier allow the moral points about fear. More on this below.

invoke and address problems of scale: skirting around the problems of full-scale reproducibility, but at the same time creating potential objections about whether or how phenomena actually do scale up or down. To take an example from Seneca,

> hic fieri illo quoque modo potest, ut inclusus aer cava nube et motu ipso extenuatus diffundatur; deinde, dum maiorem sibi locum quaerit, a quibus involutus est sonum patitur. quid autem? non quemadmodum illisae inter se manus plausum edunt, sic illisarum inter se nubium sonus potest esse, magnus, quia magna concurrunt?
>
> [Thunder] can happen this way: air, expanded by its own motion while in the hollow of a cloud, is disbursed. Then, as it seeks more room for itself it causes a sound to come from the container in which the air is held. *What?!*—In the same way that hands struck against each other make a clapping sound, can clouds struck against each other not make a sound too, all the louder because larger things come together?[22]

A little later Seneca clarifies the force of the two *potests* here. It is not that the explanation *may* hold, but that sound is not always produced, and he adds the qualification that sound can come about only when the clouds are properly arranged in relation to each other: *aversas inter se manus collide, non plaudent*, "strike the backs of your hands together; they do not make a clapping sound."[23]

In the first observation, the simple observation that clapping hands make a sound, Seneca makes an explicit link between thunder and hands: *quemadmodum*, "in just the same way." In the second observation, that of the backs of hands, the *quemadmodum* is rhetorically suppressed, but its logical force still underlies the example. But the second instance is given more immediacy, when Seneca turns an imperative verb on the reader: *aversus inter se manus collide*, "go and try it," he effectively says to the reader. The challenge is an interesting one, insofar as it implicates the reader directly in the argument, even if Seneca does not really expect us to put the book down and put backhanded clapping to the test. (Indeed, should we actually try it, his argument is weakened: it does make quite a bit of sound.)

It is worth introducing a subtle distinction here: is it the case that the non-thunder-producing contact of some clouds is being modeled in an anal-

22. *NQ* II.27.4.
23. *NQ* II.28.1.

ogy to backhanded clapping, or is it that the thesis that "arrangement of parts matters" explains the example? This is a fine point to be sure, but an important one. It is the difference between saying "We are already familiar with this theory from [some illustration]," and saying "Here is a previously unexplained phenomenon I can now explain with my new theory." Parallelism and the blatant unfamiliarity of the example lean us toward a suppressed *quemadmodum* in this instance, but not always. Consider Lucretius' treatment of dust motes. He introduces his dust motes in the context of his argument that atoms must be careening incessantly through the void. Lucretius rather deliberately flags the observability of the example: *cuius, uti memoro, rei simulacrum et imago ante oculos semper nobis versatur et instat,* "It strikes me that there is an image and reflection of this standing and waving before our very eyes."[24] When we stand in a dark room near a bright window, we see tiny specks dancing and moving in all directions, tiny specks surrounded by nothingness. That this is a model observation is clear from Lucretius' explicit statement: *dumtaxat rerum magnarum parva potest res exemplare dare,* "In this matter a small thing can furnish an example of great things" (although "size" here is metaphysical: the small thing observed is actually physically larger than the "great things," the atoms, that it models). But look at what Lucretius does next. He tells us to pay especial attention to this example for another reason: *quod tales turbae motus quoque materiai significant clandestinos caecosque subesse,* "because all this commotion shows that, underlying it, there are also hidden and invisible motions of matter."[25] Not only are the random motions of dust particles in space *like* the random motions of atoms, they are also *caused by* the random motions of atoms. It has in one respect the same argumentative form as Seneca's clapped hands, that of a model observation, and in another respect the same force as the modern phenomenon of the so-called Brownian motion, that of an easily observable effect of atomic motion. (The power of this second instance can be considerable: finally offering a theory to explain Brownian motion was one of the three coups of Einstein's *annus mirabilis* of 1905.)[26]

The use of model observations works rhetorically and logically for several reasons. One is that some phenomena, such as thunder, happen on too large a scale to be reproducible on demand by human observers, and thus

24. Lucretius, II.112–13.

25. Lucretius, II.127–28.

26. For an important qualification on just how "direct" this observation was, see Achinstein 2002.

the clapping of hands serves as an easy model for examination. On the other hand, in proposing a model observation, Seneca is open to the simple objection that clouds may not be like hands at all—who ever said they were? And notice how Seneca tries to smooth this problem out with an explicit scaling statement at the end of the first example: "all the louder because larger things come together." Inwood has argued that Seneca's frequent use of argument-by-example, being open to such simple objections, shows a commitment on Seneca's part if not to multiple explanation à la Epicurus, at least to flexible or provisional explanation.[27] This points to a deep awareness on Seneca's part of the limits of our knowledge of the physical world. I agree that Seneca is acutely aware of these limits, but I think that the argumentative force of the examples is stronger than Inwood suggests.

The crucial point here is that the rhetorical effectiveness of the model observation works only because of the background philosophical context in which readers are more or less likely to simply accept as given the idea that thunder is contact-generated. Classification plays a big part: the very use of this particular fact depends critically on what kind of thing the model observation will be seen by the reader to be, vis à vis the phenomenon being modeled. Thus the scaling problem is only really a scaling problem once the kinds are accepted as relatable. Closely related to scale is a problem of accessibility. Even if we could, for example, create clouds and blow them together, we would still be faced with the obstacle that we cannot actually get up into the sky to watch what happens at close range. As with scale, the model observation solves an accessibility problem in the same instance as it potentially raises fatal objections: do clapping hands actually access thunder? Again, the collectively assumed kind-relationship allows the example to underpin the superstructure of the case being made. There is good evidence, though, that the accessibility problem was taken to qualify, at least tacitly, the degree of certainty attainable in such investigation. At one point Seneca mocks Posidonius for pretending to too much certainty about the process of the creation of hail in clouds: *affirmabit tamquam interfuerit*, "swearing to it as though he had been there."[28] Here the language of eyewitness confirmation (*affirmatio*) stands in contrast to the impossibility of actual observation.

To see the importance of these kind-relationships, we need only call to mind that model observations are sometimes far too stark to do any real work by themselves. They don't actually tie in to the overall theory of which

27. Inwood 2005, 168–70.
28. *NQ* IVb.3.2.

Seneca is hoping to persuade his reader. That hands clapping make a sound forward but not backward is, by itself, neither here nor there in terms of Stoic doctrine—but is this always the case? If we look at what facts actually get invoked by a given theory, we find that rival schools do sometimes have their own preferences when it comes to observable phenomena. The most obvious reason for this is that some facts are simply more easily fitted to some theories than to others. Investigators, ancient and modern alike, must always make decisions about which data to accept and which data to reject (although some psychological studies also emphasize the degree to which scientists simply ignore anomalous data).[29] Such selectivity plays an important part in the elaboration of a theory's explanatory range, as well as being a tool for the clarification of a theory using the vividness of its concrete instantiations.[30] As Ian Hacking is reputed to have once said: "Philosophy is the art of the good example." Consider the phenomenon of compression: it is very cleanly explained by atomism (the atoms are simply being forced closer together; nothing is compressed per se except for the void that the atoms inhabit). But compression is not so easily explained by plenists except through the invocation of a series of further more or less complex (and more or less ad hoc) arguments about the nature of matter itself. Or look back at Lucretius' dust motes, which are a very good example for atomism. Indeed, no Stoic, Peripatetic, or Academic discusses the phenomenon anywhere, to my knowledge. Whether this is because atomism's opponents are simply too busy with other things to spend much time on it, or whether it is being more deliberately ignored ("passed over in silence" by atomism's opponents) we cannot say. But the observation itself, as it stands in the text, has a system context that cannot be ignored. It is here an entangled, rather than an atomic, fact.

Similarly if we look for other uses of *quemadmodum* examples in the *NQ*, we find some that are much closer to the Stoic heart, that are more partisan than they may at first seem. In trying to convince us that he has a ready-to-hand explanation for why water behaves the way it does—changing from air in clouds, filling and then depleting springs over the course of a day, rivers running wild one season and dry the next—Seneca offers a series of model observations, *quemadmodum* examples. That he sees them as something like judicial precedents may be indicated by his use of the phrase *iura naturae* to describe them. In any case, all the parallels make the basic point

29. See, e.g., Chinn and Brewer 1993. One might also point to Quine and Ullian 1970 for a philosophical perspective.

30. See, e.g., Thagard 1992, 82f., 143f.

of cyclical change across a wide range of phenomena: quartan fevers, gout, menstruation, seasons, solstices, and equinoxes. Simple enough, perhaps, but not innocent. Nor are they lightly dismissible: cyclicality is one of the key distinguishing features of Stoic physics, and change from one element into another is at the heart of the overall conflagration and rebirth of the cosmos. Even the fact of rain coming from clouds is brought round to this end, insofar as Seneca denies that cloud is moisture, insisting instead that it is air which changes to water in the critical moment of raining.[31]

To be sure, Lucretius also talks rather frequently of cycles—birth and death, the seasons, and so on—but these are handled very differently. In Lucretius we see an emphasis not on cyclical recurrence, but on the complementarity of generation and destruction. The centrality of the theme of growth and destruction is with us from the very outset of the poem, where Lucretius leaves Venus and Mars in their scene-setting embrace, his head in her lap, she leaning over him. Growth and diminution are nearly always handled together by Lucretius, and this for several reasons. On the atomic understanding, the growth of one thing is simply the addition to that thing of particles from something else, now made smaller (a fact for which Lucretius appeals to observation, calling on the eyes three times in four lines: *videmus . . . ex oculis nostris . . . videatur*).[32] Finally, there is a certain *quemadmodum* lurking in the shadows. The idea of generation is never very far from the example of seeds in the poem—small things that famously make big things, but that ultimately come materially from other things themselves. Nor is it very far from death. Only a few lines into the poem (*pace* editorial problems) we get the explicit connections lined up:

> nam tibi de summa caeli ratione deumque
> disserere incipiam et rerum primordia pandam,
> unde omnis natura creet res auctet alatque
> quove eadem rursum natura perempta resolvat,
> quae nos materiem et genitalia corpora rebus
> reddunda in ratione vocare et semina rerum
> appellare suemus . . .

> Now I will begin to give to you the complete explanation of the heavens and the gods, disclosing the first-beginnings from which nature creates, increases, and nourishes all things and into which she dissolves them

31. II.26; compare also II.15.
32. II.68f.

> again at death. When speaking precisely we usually call these the material or the generative parts of things, or else the seeds of things . . .[33]

Birth, growth, dissolution, destruction, all parsed in terms of underlying material entities and the language of biological generation: *genitalia, semina*. But it is not, as in Seneca, cyclicality that is at the heart of the example; it is assembly and disassembly, big things from small, one thing shuffling parts with another.

The effectiveness of these strategies of argument by comparison where the comparator is deliberately and suggestively rich, is a commonplace in all branches of ancient rhetoric: this is the ubiquitous ancient rhetorical figure called *similitudo*. In the rhetorical context, the epistemological effectiveness depends entirely on whether the speaker and the audience can agree that the two halves of the statement are indeed similar (and specific formulae can be employed to promote that impression).[34] Here shared sets of assumptions combine with stocks of established, common, or traditional similes in order to make individual comparisons more or less sound. Where comparisons recur in many contexts (thunder and clapping), the comparison is on the face of it unproblematic, even when carried over to new contexts. I have called Seneca's comparisons in the *NQ* "model observations" to signal that in bringing difficult or inaccessible phenomena closer, it is mundane, if highly associative, observable phenomena that he turns to. But the fact of observability has a powerful epistemological role in the overall argument, not just from this empirical point of view (try it and see!), but also from a significantly broader perspective. In the rhetorical texts, the epistemology of similarity as an argumentative strategy in general is conceived in concert with the certainty that is directly provided by vision. Similarity works, we are told, because the listener "sees" the comparator vividly. As in Seneca, similarity can appeal to observations that we've actually had, to observations that we plausibly could be having at any old time (backhanded clapping), or even to plausible but at present inaccessible observations, as in the *Rhetorica ad Herennium*, when an advocate asks the audience to envisage "a lute player coming on stage, magnificently dressed."[35] The eyes are stressed again and again in the theoretical discussions of argument by *similitudo*: eyes, pictures, images, light.[36] It is not just that vision and vis-

33. I.54–60.
34. *Ad Her.* IV.61, e.g.
35. *Ad Her.* IV.60.
36. Quintilian, V.xi.22f.; VI.iii.72f.; *Ad Her.* IV.61.

ibility underscore the comparisons that Seneca employs, but that vision and visibility underscore the entire epistemological force of comparison itself: Comparisons are convincing because they are visual. Seeing does big work.[37]

Observational Selectivity

To see how selectivity works, it is worth looking in some detail at how two rival authors approach the same subset of examples, paying attention to what observations they both agree on and what observations one or the other of them dodges or discredits.

Having framed his subject in the *exordium,* Seneca opens the actual argument in book 2 of the *Natural Questions* with a statement of the agreed facts (this is a standard *narratio,* to use the rhetorical terminology), and he flags the first "prior given" (*primum praesumendum*) about the nature of air on which his argument will hang.[38] Flagging this fact as a prior given is an attempt to place it beyond judicial doubt. The rhetorician Quintilian parses the technical definition of the word thus: *praesumptum, quo significatur de quo liquet,* "prior given, which refers to that which is uncontested."[39] Seneca's uncontested fact? *inter ea corpora quibus unitas est aera,* "air is among the class of bodies that are unitary." Having stated the general principle, he goes on to cite a handful of facts about air, and in many instances the observations that demonstrate these facts: it connects heaven and earth; it carries the *vis siderum* to earth (II.4.1). It has *intentio* (II.6.3) which can be seen when it sometimes picks up trees and other large objects; voice, he says, is an *intentio aeris,* and carries especially well when the air is agitated. He discusses singing, horns, pipes, water organs (II.6.5). Finally, seeds and roots split rocks, and this is also just an *intentio spiritus.*

His choice of topics is not accidental, but is aimed specifically at the evidence already on offer in support of a rival position: this same set of examples, rocks, trees, roots, and sounds had earlier been put to a different use by Lucretius, pushing not unity, but porosity (I.348–55). Lucretius reminds us that water permeates rock even into the deepest caves, that food and moisture spread throughout the whole tree from its roots on up, and that sound carries through walls, all showing that there must be voids in appar-

37. See also Goldhill 2001b

38. Seneca, *NQ* II.2.1. On the *exordium,* see Quintilian, IV.12–15; Cicero, *De part. or.* 35.

39. Quintilian, III.vi.35. I translate *liqueo* as "uncontested" to try to maintain the sense of judicial clarity implied by the verb in a rhetorical context.

ently solid bodies. That Seneca is reinterpreting for his reader the evidence earlier offered by Lucretius in these passages is further underscored by Seneca's turning now to the same argument Lucretius himself had next raised: that the movement of birds and fishes through air and water is supposed to demonstrate the existence of void. Rhetorically, the strategies of presentation are also identical. Each begins with the errors of the other school, Lucretius with what "some people contrive," *quidam fingunt,* Seneca with what "some people mangle and confound," *quidam aera discerpunt et in particulas diducunt, ita ut illi inane permisceant,*[40] which, although it is a literally accurate description of atomist doctrine, is also a cleverly damning one.

What is doubly noticeable, though, is that Seneca chooses to ignore one of Lucretius' star facts. Right in the middle of this discussion (I.358f.), Lucretius calls on heavy and light objects as evidence for atomism—how else can we explain varying weights of identically sized objects unless we suppose that some things have more void in them, some less? This is not a fatal problem for Seneca, but, to be fair, the cleanness of the atomistic explanation leans heavily in Lucretius' favor, and Seneca accordingly passes weight by.

At this point, Lucretius had backed away with a bow to his dedicatee Memmius, saying that he could offer many more proofs, but these will be sufficient for his purposes (I.398f.). Seneca, however, still wants to shore up his position, so he does not let go quite yet. To drive home the tension inherent in air, he calls on more observations (II.8.1f.): wind (again), the motion of sight, sound, and light as instantaneous, the force of air pressure on jets of water in the amphitheater, boats floating, the skipping of stones on water, and (calling one of Lucretius' observations out) the fact that sound travels through walls. This last move is a very clever one. Even as early as Strato of Lampsacus in the early third century BC, some Peripatetics had made concessions to atomism by acknowledging that the movement of sound through solid objects was a good argument for supposing the existence of at least small pockets of void in everything.[41] Seneca here reclaims the observation as an argument for an air-infused plenum, since the motion of sound would be impossible without a radical contiguity of air around and through the wall itself. Sound travels not because there are little spaces for it to travel through, but because of the tension and ubiquitous continu-

40. Lucretius I.371; Seneca, *NQ* II.7.1.

41. Berryman 1997, forthcoming (b); Lehoux 1999.

ity of the element that carries it, the Stoic *pneuma*.[42] This is a very clever reworking of evidence that would look, prima facie at least, quite damning to Seneca's side. In turning the observation around, Seneca has not merely negated its unsavory implication in the hands of the atomists, but coopted some of his opponent's best evidence to make his own case.

Something else sometimes happens to model observations which changes the precise characterization of their role in the overall argument. A good example is found in Seneca's development of his version of the nature of lightning. After the traditional survey of the opinions of the main authorities, Seneca begins his own argument effectively de novo: he is going to proceed not by engaging these authorities, but by beginning from simple observations: *dimissis nunc praeceptoribus nostris incipimus per nos moveri, et a confessis transimus ad dubia.* "Now, dismissing our teachers we begin to take our own course, and turn our attention from what is agreed to what is undecided."[43] He then launches into a classic judicial *narratio* by listing a handful of observations that are agreed by everyone (he emphasizes the formula by saying that these observations are *in confesso*). That lightning is fire is proved by appeal to observational evidence: it causes conflagrations, blackens objects, leaves a sulfurous smell. Having touched ground with mundane fire, he then moves to a discussion of its familiar causes, which are (a) percussion and (b) friction. (And note that these are causes in the simple sense of referring to the actions humans undertake when they need to cause the existence of fire, which makes them quasi–observation statements rather than what we normally mean by *causal statements* strictly construed). Once isolated, these mundane causes are immediately transposed to the clouds. We have argued from ground-level everyday observations about fire, via the (observationally established) corollary that lightning is a kind of fire, to the conclusion about how lightning is caused in the clouds. But here the transitivity is stronger than in the model observations canvassed earlier, running on more clearly spelled-out logical lines:

(a) Fire has properties *x*, *y*, and *z* (behaviors, causes)
(b) Lightning is fire
therefore (c) Lightning has properties *x*, *y*, and *z*.

The argument's strength lies in the explicit spelling out of premise (b), in contrast to model observations, where some version of (b) is assumed or

42. *NQ* II.9.4.
43. *NQ* II.21.1.

provisionally held but not explicitly laid out. And note in this instance that (b) is proved by recourse to observation. Thus the logical form stands in contrast to the rhetorical method of *similitudo* used elsewhere in the *Natural Questions*: we are here arguing more clearly *ex pede Herculeum.* To bring us quickly back to rhetoric, though, we need only look at the objection by the imaginary interlocutor at II.24: how can fire come down from above, when it naturally wants to move up? In answer Seneca recalls a favorite observation of his as a *similitudo,* one that he has used twice already in different contexts: that water can be made to go upward—against its natural motion—if force is applied.[44] What is it that does the forcing exactly? Again it is that quintessential Stoic agent, the *intentio aeris,* that powers the dramatic public display of the skyward-shooting fountain.

We have, then, two phenomena interacting: the selectivity of observational evidence, and its interpretation. Some facts fit some schools better (are system-preferred), some are part of the common canon that everyone needs to deal with (more on these in chapter 8). And sometimes, as we saw above, what looks like good system-preferred evidence can be coopted and reinterpreted by rival schools to great effect.

Examination of Witnesses

What of unusual observation claims, single-event claims, observations that cannot simply be assumed as accessible to the reader or as part of their stock of life experiences?[45] Here we must call on witnesses, those who claim to have observed some event firsthand—typically something remarkable. For Seneca, the answer lies once again in the courtroom: he treats observers as though they were judicial witnesses. Eyewitnesses thus offer testimony, as for example Euthymenes of Massilia, who claims to have seen the source of the Nile while on a sailing excursion on the Atlantic Ocean. The Nile's water, he says, is pushed inland from the Atlantic. The evidence? Euthymenes observed that the water in one region of the Atlantic that he had visited was fresh to the taste, and it had the same beasts living in it as the Nile does.[46] Like any testimony, his is wide open to corroboration or falsification. In this instance, Seneca tells us, it is contradicted by a *testium turba,* a "whole crowd of witnesses," their numbers obviating

44. *NQ* II.24.2, II.9.2, II.16.1.
45. Compare Dear 1995.
46. Seneca, *NQ* IVa.2.22.

the need to call them individually even while ruling Euthymenes' evidence out of court.[47]

It is not that we simply trust or do not trust individual witnesses. It is that, in a cultural context where judicial rhetoric is one of the main cornerstones of the education and cultural idiom of author and reader alike, everyone with an interest in the argument can be expected to be acutely aware of both the strengths and weaknesses of witnesses as judicial evidence. The strength an eyewitness can lend to a case may be considerable, but only if well-known potential objections to that particular witness are carefully circumnavigated first. Hence Quintilian's proverbializing *maximus tamen patronis circa testimonia sudor est*, "testimony is the biggest pain for the advocate."[48] By contrast, the general observational claim (the claim rooted in what we all have seen or could see if we went to the trouble) is rhetorically very different, requiring very different (and riskier) strategies to impugn it.

We see this emerge, for example, in Seneca's assertion that islands can be raised or lowered by the movement of the earth. First Seneca calls two witnesses, Thucydides, who affirms that during the Peloponnesian war the island of Atalanta was submerged for a time, and Posidonius, whose report about Sidon we are told (in the imperative) to believe: *idem Sidone accidisse Posidonio crede.* But then Seneca adds that the fact does not rest on testimony, we have our own experiences to demonstrate it: *nec ad hoc testibus opus est, meminimus enim,* "In any case there is no need for witnesses, for we ourselves remember."[49] Look, too, at the one precisely dated event we get in the entire *Natural Questions,* that of the earthquake in Campania on the Nones of February in the consulship of Regulus and Verginius. The event was so recent and immediate that Seneca tells us that he heard of it just as he was writing the *praefatio* to book 6. Does he seize on this opportunity to find witnesses? Quite the contrary. For the event, he tells us later, "has filled the world with stories (*fabulae*)," *implevit fabulis orbem.*[50] Nevertheless he does make an important exception, for one report of an otherwise unknown phenomenon that happened during the Campania quake. In offering it as evidence for his argument, Seneca becomes extremely judicial. He first carefully frames the report in legal jargon: he judges (*iudico*) the inves-

47. Compare also *NQ* VII.16.2: *praeter illum* [sc. *Ephorum*] *nemo tradidit,* "no one but [Ephorus] reports this." In this instance, Ephorus' reliability has already been impugned ("he is often deceived, often deceiving"), so we know we cannot trust him (more on whom below).

48. Quintilian, V.vii.1.

49. *NQ* VI.24.6.

50. *NQ* VI.25.3.

tigation (*cognitum*) to be worth preserving, because it is given *ab erudissimo et gravissimo viro*, "by a most learned and eminent man," praising his reliability at the very outset following standard courtroom practice.[51] And Seneca's language here is extraordinarily precise: he does not report what the man saw, but what the man *affirms* that he saw: *vidisse se affirmat*, or even more carefully a little later, what Seneca *heard him say* that he saw, *eundem audivi narrantem vidisse se* (!).[52]

We need to be very precise here. The care he is taking is not because the testimony of witnesses is in itself epistemologically weak. It is just that eyewitness testimony is going to be handled in very sophisticated ways by both Seneca and his readership. In physics as in forensics, the main trick lies in convincing everyone that the witness is indeed a good one. Here we need to avoid the all too easy trap of reading ancient rhetorical handbooks as lists of essentially dishonest strategies for overinflating the reliability of one's own poor witnesses or denigrating the reliability of the opposition's honest witnesses. The familiar hostile charge that rhetoric was the art of making the weaker argument appear the stronger has played into this perception. Lloyd has shown that even in Greek times the charges against rhetoric by philosophers and others were more bluster than substance, and had more to do with marking disciplinary boundaries than with significant differences in approach.[53] Nevertheless, the charge could be made, whereas in the Roman context, rhetoric and philosophy had come to develop not only a less antagonistic but, I would argue, a positively more fruitful relationship with each other. This is not to say that Romans failed to recognize that rhetoric could be misused, however (see e.g., Quintilian, II.xx).

By the nature of their subject matter rhetorical handbooks may suppress this to some extent, but they are the main locus of ancient discussions of the logic of probability (nonquantified).[54] This is largely because on a superficial reading they offer primarily strategies of argument rather than explicit criteria for judgment. Nevertheless a good deal can be inferred about the standards and criteria for judgment from the strategies of argument offered. Moreover the strategies of argument are not secret: everybody who matters knows the rulebook. Everyone could, in principle at least, know the range of tactics employable by both sides, plus the strengths and weaknesses of those tactics, toward the determination of plausibility, where that determination

51. Compare Quintilian, V.vii.3; V.vii.8.

52. *NQ* VI.31.3.

53. Lloyd 1990.

54. For the emergence of quantified probability, see Hacking 1990; 1975; Franklin 2001.

ultimately rests with a third side, the judges, who are equally aware of the rules and tricks of the engagement. So, for example, the reliability of a witness is not a true-or-false prior given that is to be manipulated by cunning lawyers. Reliability emerges as a consequence of the push/pull generated by those competing advocates—it is a dynamically established quantity. It is only at the end of examination and cross-examination that the judges choose to accept or not accept all or some of a witness's testimony, and only at the end of the trial that they officially pronounce on the balance of reliabilities of the various kinds of evidence. All three sides—judges, advocate pro, and advocate con—know the rules of engagement: the judges (who include the readers of published speeches) are neither gullible nor naïve. Similarly, the use of witnesses in philosophical argument is not a simple matter of blanket acceptability or nonacceptability. It is part of the larger process of persuasion, and, as in any other part of the argument, it is incumbent on authors to be careful not to leave themselves exposed unnecessarily to any of the standard lines of attack. Hence the buttressing of the moral character of the witness, the assertions of plausibility, the demonstrations of fit with the overall account.

New experiences come into the discourse of the sciences in two ways, either born general (as when many people witness a comet or a new and unusual animal goes on public display) or else born particular, as when discreet events unfold one-time-only before the eyes of individual witnesses, and Seneca ends book 7 of the *Natural Questions* with an encomium to all the wonderful as yet undiscovered things still waiting to be seen for the first time.[55] It is wrong to say that under a premodern epistemology "reports of singular events . . . had no philosophical standing."[56] They had a good deal of philosophical standing, but they were handled in special—and highly professionalized—ways, with close and explicit attention to the characters of the witnesses themselves, whose reliability needed to be assessed in terms of their individual *fides,* but also in terms of the plausibility of the phenomenon reported, as well as in terms of anticipated objections. Indeed, Seneca uses just these prongs in his hostile examination of an unusual comet observation reported by the historian Ephorus.[57] Ephorus is *non . . . religiosissi-*

55. *NQ* VII.30.5.

56. Dear 1995, 6. To be sure, Dear limits this particular sentence to the seventeenth century specifically, but the overall thrust of his argument, that something radical about the nature of experience changed in this period, shows that he has a much broader sweep in mind. Indeed, if he did not, he would be merely documenting a temporary seventeenth-century aberration in the acceptability of event-claims.

57. *NQ* VII.16.

mae fidei, saepe decipitur, saepius decipit, "not the most accurate witness, often deceived, more often deceiving" (not least, because he is a historian, jokes Seneca). In any case, *quis enim posset observare?* "who would even be capable of observing [such a thing]?" Ephorus is utterly unreliable, the observation impossible. That these categories are being conceived judicially is apparent from the introduction to the passage, *contra argumenta dictum est, contra testes dicendum est,* "The evidence has been refuted, now the witnesses shall be," a formula that hearkens back to the *Rhetorica ad Herennium,* both in its wording and in Seneca's implementation of it.[58]

As both C. A. J. Coady and Steven Shapin have pointed out so forcefully, we have an almost overwhelming—and overwhelmingly duplicitous—tendency to think that what we hear from other people is somehow less worthy of trust than what we think we have seen for ourselves. We will see in chapter 6 how the category of "what we think we have seen for ourselves" has its own epistemological problems. I want here to propose that part of the reason that category is overstretched is that we are attempting to use the observation to paper over some of the knowledge that we very much want to—nay, *must*—keep, even if the degree of certainty we have belies our lack of certainty in its source, *viz.* testimony. This puts us in a kind of epistemological denial: the fact is, most of what we know we have taken directly or indirectly on trust.[59]

What are we to make, now, of Coady's claim early on in his book that testimony was not treated as part of knowledge in antiquity? "Amongst the ancients and medievals, the view of knowledge as a kind of thoroughgoing rational understanding militated against treating beliefs based on testimony as part of knowledge."[60] He is right, but we need to be very careful how we read him. Coady is effectively using two different definitions of the word *knowledge* here, and we would do best not to conflate them.[61] The problem is that what the ancients call knowledge in this sense does not correspond to what we normally call knowledge, not even in many epistemological contexts (I suspect that much confusion could be avoided if we simply refused to translate the Greek ἐπιστήμη). Most ancient epistemology has extraordinarily high standards for what will count as capital-K Knowledge.

58. *Ad Her.* II.9.

59. Coady 1992; Shapin 1994, chap. 1. See also Audi 1997; Lipton 1998; Kusch 2002; Thagard 2006.

60. Coady 1992, 6.

61. Note that one school, that of empiricist medicine, took a radically nonsceptical stance with respect to testimony, counting reported experience as personal experience, and therefore knowledge. Testimony, in short, equals observation.

On the other hand, when we talk about knowledge now (especially in the wake of Coady's work) we find all kinds of discussion of the acceptability of testimony. Essentially, the ancient schools draw the lines between knowledge, persuasion, probability, and plausibility differently than we do, but this does not mean that they do not have any way of dealing with persuasion and probability, simply because they do not define those as knowledge, strictly speaking, and handle them under the heading of rhetoric rather than epistemology.

The Natural Authority of Morals

In judicial practice, the most important factors contributing to the reliability of testimony are the reliability of the witness and the plausibility of the account itself, both prima facie and in connection with other evidence. But the authority of individual witnesses *as witnesses* is ultimately not simply a blanket question of their reliability, but a question of their individual moral character. Particularized moral judgments are passed on individuals. This person is virtuous, therefore reliable, that person is vicious, therefore unreliable. Class distinctions and other institutional considerations surely play roles in those moral adjudications, but the presentation also often cites particulars: so-and-so who did such-and-such. Here the citation of moral qualities and specific past moral behavior by the individual serves to buttress the degrees of trust in whatever they are reporting. The rhetorical handbooks are unanimous on this point: the easiest way to malign a witness is to show that his manner of life is vicious.[62] This use of individual character and of that individual's past acts in the establishment of the reliability of the individual witness has a curious side effect. It means that the incorporation of testimony into the scientific canon as general truth relies not on general principles about the reliability of witnessing in and of itself, but on moral particulars about individuals.

There is a corollary to this: witnessing itself is a moral act. Observation is not just theory-laden, it is ethically laden. Thunder, Seneca tells us, sometimes makes a violent popping sound, *quale audire solemus cum super caput alicuius dirupta vesica est,* "as we are used to hearing when a bladder is popped over someone's head."[63] Both Pliny and Lucretius use popping bladders as a comparator for thunder,[64] but there is an added nuance

62. *Ad Her.* II.vi.9; Cicero, *De part. or.* 49; Quintilian, V.vii.26.

63. *NQ* II.27.3.

64. Lucretius, VI.130; Pliny, NH II.113.

in Seneca in the words *super caput alicuius.* We might legitimately wonder what on earth the qualification "over someone's head" is meant to convey. Neither Pliny nor Lucretius, for example, says that their balloons were popping as part of a practical joke. In fact there is an important ethical point to this image of juvenile behavior that Seneca draws out at some length: that when people hear very loud thunder they sometimes fall to the ground in a swoon, occasionally even with lasting derangement. It is not just the physics of the popped balloon that matters; it is also the fear. Here Seneca's model observation is polyvalent. It makes the physical point of the familiarity of the mechanism at the same time as it makes a fundamental moral point about the fear, a moral point that he will use to tie the whole of the *Natural Questions* together at the end of book 2: if we do not fear death, we will not fear the causes of death.[65] And in any case, death is inevitable, so merely delaying it should not be cause for great concern. As Seneca puts it in the wonderful peroration to the work as a whole, *nemo umquam timuit fulmen, nisi quod effugit,* "the only people who fear lightning are those whom it has never struck."[66]

But the point is not merely made for the reader on sober reflection about the example. There is an immediacy to it (dare I say, a visual immediacy) in the suppressed but still strongly present image of the reaction of the victim to the popped bladder. His reaction is completely risible from the point of view of the onlookers, onlookers among whom the reader is numbered (*solemus*). So, too, from another point of view, is the swooning of those who fear thunder. Here the bladder does not just model thunder, but also domesticates it—thunder becomes less terrifying when seen in light of the bladder.[67] This is not just because we understand the physics of it, but also because we can see how silly we must look when seen from another point of view.

The use of the risible domesticating model is likewise put to spectacular effect by Lucretius at the end of book 4. He begins the book with a long explanation of the causes of vision. From there he moves to the other senses and motor functions, and then finally to dreams, whose causes lie in the same images (*simulacra*) that enabled vision in the first place. But *simulacra* have a further effect: they excite desire, and desire is an ethical problem. In the description of human copulation that follows, Lucretius keeps Venus wedded very much to Mars: orgasm is likened to death from

65. I am following Codoñer Marino 1979 in ordering the books. See also Hine 1996, xxiif.
66. *NQ* II.59.13.
67. On domestication in this sense, see Williams 2005.

a battle wound, passionate kissing and violent biting are faces of the same coin; the *burning* of desire is emphasized. So, too, is its visual nature: we crave through the eye; we want more and more of those images. Desire also perverts vision, causing us to falsely see beauty where none is present and to invent euphemisms in order to, as the joke about computer software goes, present the bugs as though they were features.[68] But the desire is—subjectively speaking—very real and powerful. Even the overwhelming immediacy of the feeling should not carry us away, for pain and enslavement are never far off. Do you not see, he asks, the torment of the bond? And here he delivers the *coup de grace*:

> nonne vides etiam quos mutua saepe voluptas
> vinxit, ut in vinclis communibus excrucientur?
> in triviis cum saepe canes, discedere aventes
> diversi cupide summis ex viribus tendunt,
> cum interea validis Veneris compagibus haerent.

> Do you not see then how those bound by desire often suffer in their mutual bondage? So dogs at a crossroads, straining to separate, often lustily pull in different directions with all their strength, all the while held together with the chains of Venus.[69]

The effect is a powerful one. Like Seneca's popped bladder, the effect implicates the reader-as-observer ("do *you* not see?"). The humor and the revulsion are instantaneous and coeval. With this utterly pathetic image in mind, how seriously can the reader take lust now? The terminology gets turned on itself all through this passage, the dogs pulling *lustily* apart, *summis ex viribus,* the chains of Venus powerfully and damningly reimagined. Although Epicurus seems to have handled hedonism very carefully in order to defuse any overzealous interpretations of what might be permissible or desirable, not all professed hedonists followed him in that respect, and slander was fairly cheap and easy to come by, one brush happily serving to tar all together. Lucretius uses this to his advantage by allowing desire just enough room to gain rhetorical traction at the outset of the section, but at the same time never letting it stray far from pain. The final observation in which the reader is implicated is doubly interesting, insofar as it not only governs the picture Lucretius paints of the dogs, but even more immediately governs a

68. IV.1155f.
69. IV.1201–5.

moral observation: "Do you not see then how those bound by desire often suffer in their mutual bondage?"

Seneca's big treatment of the morality of observation is similarly highly sexualized in the famous case of Hostius Quadra. Too much ink has lately been spilled on this one sad libertine for me to add to it in good conscience, apart from simply pointing out the number of times Seneca flags inappropriate vision as being at the center of the moral problem with Hostius.[70] The interesting rhetorical feature is that, in contrast to Lucretius' dogs, Seneca does not lure us in through an agreement on the pleasures of love, but from the start calls on our moral outrage at Hostius' perversions (he seems to have had a fetish for funhouse mirrors). All of this ties in with Seneca's constant refrain about the ills of decadence, one that trades on and reinforces the reader's moral prejudices. The immediacy of our prejudice against decadence works in the same way as the immediacy of our agreement that clapped hands make a sound, such that Seneca is calling on the reader's ethical observations just as he did our physical ones.

The eyes are similarly implicated in other passages on decadence, although in a slightly different relation to the reader. In his diatribe against gourmands, Seneca focuses on their insistence on seeing a mullet die before their eyes as proof of its freshness. By the first century AD, the mullet had become a kind of byword for excess luxury, as also witnessed by Pliny, who pauses from his descriptions of fish and their habits in his *Natural History* to rail against gourmands, one of whom, Asinius Celer by name, once paid an astounding 8,000 sesterces for a single fish (an amount equal to just about nine years' salary for a legionary soldier).[71] On Seneca's account, the diner's desire to watch the mullet die before eating it is not just rooted in the belly's demand for freshness. No, it has become something even worse: *oculis quoque gulosi sunt,* he says, "they are even gluttonous with their eyes!"[72] They now lust after the sight itself of the dying fish. But here an interlocutory voice begs to differ: "Nothing," it says, "is as beautiful as a dying mullet."[73] The use of the interlocutor to voice what is clearly the wrong side of the moral divide here is interesting but not entirely unexpected. If the shift in moral perspective that so effectively underpins the popping bladders

70. The tale of Hostius Quadra has been much discussed from many angles. When seeing enters the discussion, however, it is invariably focused only on the nature of mirrors and reflection of *the self* rather than on seeing per se. See, e.g., Johnsson 1995; Berno 2003; Toohey 2004, chap. 8; Bartsch 2006.

71. Pliny, *NH* IX.67.

72. *NQ* III.18.7.

73. *NQ* III.17.2–3.

incident allows us as onlookers to see the silliness of our own behavior and the various snares attending observation, it is notable that such an argument does not aim for a more sanitized god's-eye view. The shift in moral perspective that we are led into in the case of the bladder popping, as with Lucretius' dogs at the crossroads, is hardly disinterested. We are deeply implicated. As laughing onlookers, we are above the immediate victims, but in no way apart from them. With mullets, we see flesh given to what appears to be a common contemporary opinion and are further carried along by the interlocutor's sensuous description of the dying fish. (Here as with Hostius Quadra, Seneca shows himself a master of what Cicero calls the "visual *subjectio*.")[74] So our moral predispositions are being drawn into the line of fire rather than playing executioner themselves, as they had in the case of Hostius Quadra.

Declamation and Certainty

These moral challenges to the reader, the use of an interlocutory voice, and the use of imperative verbs are important parts of the rhetoric of the *Natural Questions*. No extended study of these literary devices has yet been undertaken, although the general impression of a "dialectical" method is frequently remarked on.[75] Paying close attention to the declamatory aspects of the text, however, sheds some much-needed light on the issue.[76] If we compare the various voices being used in just about any published Latin judicial speech, we see that to call one of those voices *the* interlocutor is to miss the point. We need to think of Seneca as using a variety of voices and techniques to make his case, sometimes addressing the reader directly, sometimes posing hypothetical objections from imagined opponents (the rhetorical figure of *subjectio*), sometimes simply giving voice to or taking advantage of what he may expect to be the reader's own reactions. In reading published rhetorical speeches (Cicero's, for example), the use of these various voices is never a problem for the reader because of the reader's expectations about what is on the page. The genre demands that we visualize

74. On this figure, see, e.g., Cicero, *Or.* III.202; Quintilian, IX.ii.40.

75. Williams 2005 does explore the interlocutor (he assumes a unity of voice) as an embodiment of flawed vision in book 1 of the *NQ*. Also, in the introduction to the Loeb translation, Corcoran 1971, xxv, offers some remarks mostly grounded in how "confusing" he thinks Seneca's use of the interlocutor is. Harry Hine has a forthcoming paper that looks at some aspects of voice in the *NQ* and other texts. On the Latin imperative generally, see Risselada 1993.

76. On declamation as a genre in the early empire, see Bloomer 1997; Beard 1993; Gunderson 2003.

a performance in front of a judicial audience. When Cicero switches from addressing the *iudices* in one sentence to Cato in the next and Servius Sulpicius a paragraph later, the argument is easy enough to follow because we are aware of the various parties and their interests in the case and we see them sitting in the room around him as he delivers his arguments. When objections and questions get posed in one voice or another, we can see their argumentative force without much difficulty, even if (as often happens) we would be hard-pressed to say exactly whose voice Cicero is using, or which individual sitting up in the stands he may be gesturing toward. Is he aping something the prosecutor has said? Is he giving voice to common gossip? Is he simply covering himself against potential objections? No matter, the thrust of his argument is clear. Similarly with Seneca, we need not imagine a single coherent voice every time he turns outward in the argument. Sometimes he is directly addressing the reader as a *iudex;* sometimes he is voicing objections that he imagines us or others to have. As a judge, the reader is not supposed to naïvely concede everything Seneca tells us. Instead, in using the interlocutory voice Seneca is clearly signaling that he is perfectly aware that there are other sides to the case he wants to make, but that he can also meet those objections. Is he giving us the final word on thunder? No more than Cicero gives the final word on Roscius. That is to say, he presents the best case he can to our inspection, a case that is always conditioned by the presence of a (real or imagined) opposition and one that is, in the end, subject to a third party's reflection and judgment.

The method is not, in the strictest sense of the word, *dialectical.* No pretence to truth emerges from the discussion between Seneca and the interlocutory voice. Instead, the adoption of a judicial strategy in the raising and answering of these objections shows Seneca to be acutely aware that the final decision is out of his hands. It supervenes on his text in a way that he has no final control over, even if he exercises a masterful medial control within the argument itself. Where Cicero published speeches whose judicial outcome was already known, Seneca is publishing an investigation borrowing heavily on model declamations of the *contraversia* type, whose outcome is uncertain.[77] We see this even more clearly in one of Seneca's letters. Written during the same period as the *Natural Questions* and likewise addressed to Lucilius, Letter 65 picks up several threads from the *NQ.* On the one hand, it offers a philosophical justification for just the kind of physical investigation Seneca undertakes in the *Natural Questions* (after

77. It may be worth remarking here that, as an anonymous referee reminded me, Seneca's father was himself the author of a famous (and still partly extant) series of *Controversiae.*

all, what business does a serious philosopher have with earthquakes and thunder when there are such pressing ethical problems still to be solved?). And on the other hand, it explores some very abstract questions of causation, questions that are not answered in the more prosaic approach to causation found in the *NQ*. Most strikingly in the present context, though, is Seneca's very clear use of judicial models in the letter, up to and including his explicit canvassing of the theoretical constraints on judgment and certainty. He repeatedly refers to Lucilius as the *judge* over the cases being put in the letter. As in the *NQ*, we see the familiar pattern of interlocution, sometimes giving voice to the judge's doubts, sometimes to his opponents or their imagined supporters.[78] What emerges is not just more light on the rhetorical structure of the *Natural Questions*, but an explicit discussion of the epistemological implications of the judicial model in its relation to the reader:

> fer ergo, iudex, sententiam et pronuntia, quis tibi videatur verissimum dicere, non quis verissimum dicat. id enim tam supra nos est quam ipsa veritas.

> All right, then, you be judge and give a verdict; proclaim which one [of us] seems to you to say what is truest—but not which one says what actually *is* truest, for that is as far above us as truth itself.[79]

The uncertainty inherent in the form is important to our understanding of the epistemology underpinning Seneca's arguments in both texts: more Columbo than Galileo, not reading the "book of nature," but "booking" it, as the police-show slang goes.

But isn't this where cutting-edge science often sits? Cases put by researchers may eventually come into something like a consensus, but that process is never in the control of any one study. And indeed, the consensus is not typically *about* any one study. It supervenes on the arguments of many studies, qualifying them and synthesizing them into a probable (if not strictly probabilistic) picture of the world. An obstetrician, for example, may put the case to a pregnant woman for or against fetal heart monitors (which some studies find beneficial, some risky), and in doing so she is ulti-

78. Inwood's commentary on Letter 65 (2007) points to the occasional ambiguity for where a particular interjection is coming from. In one instance, it is unclear whether an objection is meant to be by someone imagined as being in the room with Seneca, or by Aristotle himself (see, e.g., commentary on 65.4).

79. Translation following Inwood 2007.

mately making a highly synthetic adjudication. She may be very confident of her final recommendation, but the epistemology rests, in the end, on a rather intangible weighing of probabilities and trust in various studies, methodologies, biases, and results. *L'hypothèse est synthèse* in more ways than Bachelard had anticipated:[80] it is synthetic across time, across individual studies, and even (at a slightly different level) across groups of scientists. Where, for example, does the certainty that humans cause global warming rest if not across multiple individuals, each singly adjudicating the results of multiple studies?[81] What we might call the truth-value of theories emerges in an *engagement* with multiple arguments and competing evidence. Seneca shows that he is perfectly aware that his argument is part of a similar epistemological engagement. As was the case with the question of the reliability of individual witnesses, truth emerges here as a dynamic quantity, subject to the pushes and pulls of competing arguments, with final arbitration in the hands of the reader.

80. Bachelard 1934, 6.

81. I am well aware that recent work in the sociology of the sciences should have me worried about using language about what any member of a professional community does *singly*, but I am here pointing to the fact that individual decisions are still being made. This is not to deny or downplay the roles of professional, institutional, or educational pressures on that decision-making process.

CHAPTER FIVE

The Embeddedness of Seeing

In the last chapter we explored some of the strategies used in Roman sources for the deployment of observation claims within arguments about nature. Building on that analysis, we might say that there are, in effect, four perspectives from which observation statements find themselves launched in our sources, each with their own strengths, entailments, and potential pitfalls. In the first instance, an author may point to general observations along the lines of "We see that . . ." or "It has often been seen . . ." General observations are particularly robust and common in our sources. So, too, an author may call on the reader's own observations, as with Seneca's appeal to backhanded clapping. The author may also draw on third-party observation reports, although as we saw, the author then needs to be careful about whose reports he is using and how he sets them up. Finally, the author can claim to have observed something himself, which he is now reporting to the reader. In many philosophical contexts the last of these four, the "I myself have observed" kind of claim, tends to be comparatively rare. In what we might call more technical contexts, however, personal observation claims can take a more prominent place.

From a rhetorical and epistemological point of view, the author's use of his own observation reports is in one way fairly straightforward, insofar as the problem he faces vis à vis the reader is just a variant on one of the overarching problems he always faces: that of convincing the reader that he is reliable on points of fact, interpretation, and argument. But there is a quiet yet insistent epistemological problem that this issue of broader reliability does not directly address. Even if we grant an observation to the author, how do we know (or how does the author himself even know) that what he is

This chapter was originally published as Lehoux 2007b.

reporting as an observation is an accurate representation of what the world is like? The question becomes more pointed the more technical an observation becomes, since the more minute, the more difficult, or the more highly skilled an observation is, the more the author's own discernment, training, and care move to the epistemological foreground. Part of what I argue in this chapter is that the questions of discernment and of how we know that sensory experience can map onto what is really out there in the world began to show a marked prominence in the historical period a couple of generations after Seneca, emerging most noticeably in a set of debates between Sceptics and two of the most prolific and sophisticated deployers of their own observations that the ancient world has preserved for us, Ptolemy and Galen. We will see that both Ptolemy and Galen adopt identical strategies for meeting the Sceptical challenge, where they both put together detailed physical explanations of how *seeing* happens in the world, explanations that for them close off any potential inroads for the Sceptics.

As we explore these debates, however, and as we see how Ptolemy and Galen separately meet the Sceptical challenges to the uses of observation in arguments about nature, we will come to a point for both authors where they make use of identical explanatory principles (or perhaps better, laws) to close off their respective physical systems of perception: in each case, this is the law of "like affects like." Both of them at this point appear satisfied that they have covered all the bases they needed to cover for a physical explanation of seeing. Nevertheless, the modern reader is left asking why they thought that like-affects-like was explanatorily sufficient in itself, without further elaboration. We shall see that their satisfaction turns on the way arguments and even proofs find themselves already embedded in larger epistemological and (as we move to later chapters) ontological contexts.

Doubts about Vision

To begin, consider the following problem:[1]

> Why is it that when we look in a mirror, left and right get reversed, but up and down do not?

Like any good puzzle, once solved it may seem surprising that the question should have posed any problems at all. And yet it can be genuinely difficult

1. For philosophical treatments of the puzzle, see Block 1974; Denyer 1994; see also the appendix to this book.

if you've never thought it through before. Perhaps the trick to the question lies in the way the answer hides itself in the space just between optics and perception, between physics and psychology, image and imagination.

Indeed, a newcomer's initial attempts to answer the question highlight an almost instinctive tendency to compartmentalize the different aspects of this single phenomenon under the auspices of different sciences: optics, physiology, psychology. In the second century AD, when Ptolemy first faced this same question in his *Optics*, he did not face the same problems I do in seeing an answer, partly because he was himself working in territory that only later would become liminalized, hiding in the boundary areas of the different modern sciences. To put it another way: Ptolemy does not have the same blind spots as I do. By this I do not mean to say that his science is more inclusive, but that it moves over different ground, and that the ways in which it does so are revealing: the concepts he takes for granted, the phenomena he seeks to explain, the tools he chooses, and the answers he is satisfied with, all these will shed light not only on Ptolemy's project, but also on the ways in which modern ideas of vision and perception are similarly embedded. Part of what I am driving at here is an idea that has been demarcated in different ways by different disciplines at different times, and aspects of it have been flagged by Hacking, Goodman, Foucault, and others:[2] Any given way of framing questions of truth and falsity about the natural world is bound up in networks of relations, and requires a background of standards, concepts, methods, tools, and objects against which truth and falsity can be judged. This background is necessarily both highly complex and, under the normal constraints of day-to-day practice, invisible.

Partly because of broad historical, intellectual, and cultural currents,[3] the second-century AD Roman empire is particularly fruitful for the sciences, and it is here that we find two of the most sophisticated, prolific, and ultimately authoritative of ancient scientific authors: Ptolemy and Galen. Each of them, as it turns out, works on very different aspects of the phenomena of seeing: Ptolemy in mathematical optics, and Galen in physiology. In pursuing their different ends, though, each of them needs to deal with the physics of vision and the mechanisms of perception.[4] The sets of questions these

2. Hacking 1983 puts these issues under the umbrella of styles of reasoning. Goodman 1978 frames it as a question of worlds. Feyerabend 1975 tries to locate it under the rubric of incommensurability. Foucault 1966 refers to epistemes, and Pocock 1985 speaks of discourses.

3. See, e.g., Kollesch 1981; von Staden 1997. On the Second Sophistic generally, see Bowersock 1974; Anderson 1993; Borg and Borg 2004; Whitmarsh 2005.

4. Throughout this chapter, the careful reader will notice that *sensation* and *perception* are being used virtually synonymously. Neither Ptolemy nor Galen had access to Bertrand Russell's

pose are particularly interesting in that, like the mirror puzzle, they run up against the boundaries between the objective and the subjective, between the physics of the world, the makeup of the human body, and the idiosyncrasies of perception.

Why Ptolemy and Galen in particular? These are not two authors who commonly find themselves stood up together for focused comparative analysis.[5] There are several reasons that I do so here. One is that Galen and Ptolemy, exact contemporaries that they are, have investigative domains that are in one sense mirror images of each other, standing on either side of the eyeball, each probing its secrets and the relationships in which it is rooted, but each from the opposite end of the pupil. Ptolemy is looking at the way visual lines move in the world at large, how they come to the eye itself, how they can be physically altered to create illusions and errors, and how such common illusions and errors of perception can be explained purely geometrically, in terms of the propagation of visual rays. His account of these phenomena is thorough, detailed, and rigorous. But he also knows that his explanation of vision is incomplete if he cannot give some account of what happens inside the eye itself to produce the phenomena of vision that we experience. Standing on the outside, he knows he must make at least some effort to look in, if he is to give a full account of seeing. Contrariwise, Galen is carefully examining the structures in and behind the eye, how it is physically constituted, how it is connected to the rest of the organs, and what these various physical connections might mean for how we actually come to perceive anything. But he also knows that his explanation of vision is incomplete if he cannot give some account of what happens outside the eye to produce the phenomena of vision that we experience. Standing on the inside, he knows he must make at least some effort to look out if he is to give a full account of seeing. It is just at this point that both authors begin to get hazy in their accounts, as they look beyond their customary spheres, from their own side of the eye to the other's.

Now why should both Ptolemy and Galen care about giving such a full account of seeing? Ptolemy's predecessors in mathematical optics, for example, show no concern with what happens to visual information once it reaches the eye. But for both Ptolemy and Galen—and this is where their different domains intersect—there are epistemological considerations that

very influential distinction between sensation and perception (see Russell 1921). Indeed, the kinds of psychological burdens that Russell is flagging in perception as interposing and imposing on sensation become especially problematized (and so begin to need systematic explanation) only with the advent of professional psychology in the nineteenth century.

5. Although there are a few exceptions to this: Cardona 1975; Manuli 1981; Long 1989.

are ultimately at the foundation of the very methodologies they are employing: If we have no account of the relationship between sensation and the mind, how can we use the data of observation as a fundamental ingredient of what counts as *knowledge*? In order to open up a secure epistemological space for perception to stand in, both Ptolemy and Galen go to some lengths to show that there are no serious or insurmountable impediments to the transmission of facts from the world at large to our minds via the senses in general, and via sight in particular. That is, if we can understand how visual impressions come from the world into the mind, and we can show that such a transmission is secure (or at least that its insecurities are comprehensible and manageable), then we have a sound epistemological base on which an empiricism can rest. This grounding is all the more crucial, I will argue, in the philosophical context in which they were both working, where several varieties of Scepticism—sometimes very radical Scepticism—were leveling nontrivial objections to the reliability of sensory experience in general.[6]

Related to this is the fact that for Galen and Ptolemy, the ways in which they frame not only their own proper investigations, but also how they probe into each other's territory will be conditioned by their intellectual resources and philosophical commitments.[7] In both cases these commitments show a high degree of philosophical independence but at the same time a readiness to use the fruits of the rich second-century intellectual milieu as both resources and grounding to solve very specific problems in the sciences, problems that only a working scientist would encounter so intimately.

Lastly, because they are the predominant figures in their respective fields in the Arabic and European Middle Ages, both Ptolemy and Galen later stand as the quintessential authority figures against which early modern science (Copernicus, Harvey) would see itself as struggling to emerge. We could tell a similar story about Aristotle's relation to Galileo in this context, but what Ptolemy and Galen share with each other is a much stronger emphasis on, and control of, careful and systematic observation than we

6. The extent to which the particular problems canvassed in this chapter predated the second century is unclear. Scepticism had been around in various forms for a long time indeed, as had empirical investigation and studies of perception. Nevertheless, we see a flurry of activity at the conflux of the three in the second century AD, as I hope to show. On Galen and Scepticism, see Hankinson 2008. Contrast Long 1989, who sees in Ptolemy a "robust silence" with regard to Scepticism.

7. Both Galen and Ptolemy are often classed by modern readers as philosophical *eclectics*. Although the word is handy, it is not uncontroversial in this particular instance, and I would certainly not want *eclectic* to be understood as pejorative. On Galen's eclecticism, see esp. Frede 1981; Hankinson 1992; Donini 1992; von Staden 2000. On Ptolemy's, see Long 1989; Smith 1996, 17–18. On eclecticism in general, see, e.g., Donini 1988.

find in Aristotle. What we have in Ptolemy and Galen are fully fleshed-out empirical methodologies, which show sophisticated attempts to control the vagaries inherent in observation through the improvement of observational techniques, through the understanding of where observational error can intrude, and through the careful minimizing of the effects of such error. On the epistemological front, they also agree on a crucial point: for them the relationship between perception, objects, and ideas is not, in the first instance, one of *representation*. It is instead one of direct physical causation.[8] The precise details of how this is so will be worked out in different ways by Ptolemy and Galen, but for each of them this line of causation is central to the grounding of experience.[9]

Mechanisms of Seeing in Antiquity

It is worth having a brief look at the ancient visual and perceptual theories that furnish the background against which Ptolemy's and Galen's investigations are framed. If we start from the eye itself and look outward, there are a number of competing accounts of how visual information comes from the world at large and into the eye. We should also keep in mind that the basic

8. Philosophers are wont to characterize representational accounts of perception in contrast to something called "common sense" realist accounts, but I think here they generally underestimate the common-sense nature of the representational accounts, which are, if anything, far more pervasive and deeply rooted (at least since Locke). See, e.g., Putnam 1999; Foster 2000.

9. Lorraine Daston and Kelley Wilder have pointed out to me that the line of argument here got a kind of second life in the nineteenth and twentieth centuries in debates about the nature of photography. These debates run on two different (if overlapping) planes: (1) Is photography representational or (merely) causal? If only causal, can it be *art?* (2) What is the nature and value of photographic evidence versus other kinds of evidence? Both of these issues put the question of human agency and mediation between object and image front and center (and situated at just the point where Scepticism would drive the wedge in for the ancient problem of perception). Much of the debate on photography (as T. Cohen 1988 pointed out very rightly) sees the camera as implausibly autonomous (see also Snyder and Allen 1975; contrast Walton 1984). Take Bazin's very influential and highly provocative framing: "L'originalité de la photographie par rapport à la peinture réside donc dans son objectivité essentielle. Aussi bien, le groupe de lentilles qui constitue l'oeil photographique substitué à l'oeil humain s'appelle-t-il précisément « l'objectif ». Pour la première fois, entre l'objet initial et sa représentation, rien ne s'interpose qu'un autre objet. Pour la première fois, une image du monde extérieur se forme automatiquement sans intervention créatrice de l'homme, selon un déterminisme rigoureux" (Bazin 1975). In philosophical aesthetics, this same line of thought led Scruton 1981 and Walton 1984 to argue that photographs (being causal) were not representations, which seemed to many to mean that they could not be art. Attempts to salvage representation in photography (e.g., Brook 1983—misreading Goodman 1968) have unfortunately sometimes relied on interpreting "representation" linguistically (far too limiting!), a trap that Scruton thankfully avoids, as does Currie 1991. Cohen 1988 is still, I think, the best way out of this whole quagmire. Here I cannot resist the urge to paraphrase the NRA: cameras do not shoot people, people shoot people.

phenomena that get ancient optics off the ground are all based around how objects in the world appear to us and why.[10] Why do objects appear smaller when farther away? Why do mirrors produce images? How do we perceive distance and shape?

Ancient optics is not about light, it is about vision. The modern idea that visual information is carried in the first instance by the action and movement of light has become so ingrained for us that it is often difficult to set this assumption aside and to allow some room for the very foreign mechanisms of sight in ancient optics. Nevertheless, the movement of light as the primary mechanism of vision is not so obviously given in the visual phenomena themselves. In fact, this claim needs to be argued for at length in our earliest source for the idea, which is the Arabic writer Ibn al-Haytham.[11] In antiquity light played some very different roles in seeing, and not every account of seeing seems to have even felt the need to invoke or explain the role of light in any detail. Perhaps the oddness of ancient light is seen most clearly in Aristotle, for whom light was nothing more than the actualization of the inherent (but passive) tendency of air to be transparent.[12] That is: air (or water) is potentially, but not always, see-through. At night, the potential transparency is unactivated, and the air is accordingly nontransparent, so we cannot see through it. Light is just the actualization of the air's potential transparency, which thus allows visual forms to pass.

This is a very foreign idea, indeed.

Turning from physics to mathematical optics, we find virtually universal agreement on a different model. Unlike the modern model, where the eye takes in light and thence information, for ancient mathematical opticians the eye instead sends out some kind of radiative visual force that contacts objects in the world and somehow then passes information back to the eye. The details of this radiation vary from writer to writer, but the basic model is one of *extramission* out from the eye, rather than *intromission* into the eye.[13] This is not to say that everyone in antiquity who writes on vision in general is an extramissionist (a good many philosophers are definitely not),[14]

10. See, e.g., Lindberg 1976; Simon 1988; Smith 1999.

11. See Sabra 1989; Smith 2001.

12. See Smith 1999; Bonadeo 2004 (esp. chap. 4); Berryman, forthcoming (a).

13. To be sure, the ways in which the radiation passes this information back do usually depend on light as an actor, and this will complicate or at least qualify the use of the word *extramission.* See Smith 1999, 23f. Note that here and throughout *radiation* is used in its root sense to refer only to things that travel in straight lines, i.e., as mathematical *rays,* rather than in the modern physical sense.

14. The Epicureans are perhaps the most obvious example, but Plato and Aristotle, like most of the Presocratics for whom we have any information, can not be counted as extramis-

but those who work in the field of mathematical optics do for some reason have a strong tendency to adopt this model. In general, radiation that travels in straight lines offers a clean way to mathematize vision, but why then choose radiation out rather than radiation in, which is presumably just as rectilinear?

In fact, there are several empirical phenomena that we know inclined some ancient authors to extramission. One is squinting: why is it that people with weak eyes can see better when they squint? On an intromissionist theory, squinting should reduce the amount of visual information able to enter the eye, and so should decrease visual acuity rather than increase it, but on the extramissionist theory, squinting has the (obvious) effect of concentrating the visual rays in a smaller space, thus enabling better vision. Here is how the Pseudo-Aristotelian *Problems* had put it, some four hundred years before Galen:

> διὰ τί οἱ μύωπες συνάγοντες τὰ βλέφαρα ὁρῶσιν; ἢ δι᾽ ἀσθένειαν τῆς ὄψεως, ὥσπερ καὶ οἱ πρὸς τὰ πόρρω τὴν χεῖρα προσάγοντες, οὕτω καὶ τὰ βλέφαρα πρὸς τὰ ἐγγὺς προστίθενται ὥσπερ χεῖρα; τοῦτο δὲ ποιοῦσιν, ἵνα ἀθροωτέρα ἡ ὄψις ἐξίῃ, δι᾽ ἐλάττονος ἐξιοῦσα, καὶ μὴ εὐθὺς ἐξ ἀναπεπταμένου ἐξιοῦσα διασπασθῇ.

> Why is it that shortsighted people squint their eyes to see? Is it because of the weakness of their eyes, as in those who bring their hands up over their eyes when looking at something far off, and so bringing the eyelids together when looking at something close is like using the hand? They do this so that the vision they send out is more bunched up (because it is exiting through a smaller aperture) and not dispersed immediately out of a wide-open aperture.[15]

For Galen, seeing is dependent on the extramission of a substance he calls psychic *pneuma,* and this extramission is manifest in the fact that when one eye is closed, the other pupil widens (to allow passage to the pneuma whose extramission has been blocked at the closed eye).[16] So also intromis-

---

sionists either. Nevertheless, Mary Beagon has pointed out to me that Democritus' explanation of the Evil Eye uses a kind of extramission of malign particles to account for the effect (something similar can be found in Plutarch, *Mor.* 680b–682b), although this is not an explanation of vision *per se.* See Beagon 2005, 139f. For a good general overview of the different optical theories on offer in antiquity, see the introduction to Smith 1999. Finally, Smith 1988 questions Lindberg's idea that there was a distinct "philosophical tradition" in Optics.

15. Pseudo-Aristotelian *Problems* 959a2f.

16. Galen, *De plac. Hipp. et Plat.* VII.4.11f.

sion is plagued in antiquity by many problems: how could the image of a mountain shrink to fit into our eye? How could the mountain shed so many images as to be viewable to any observer on any side and at any time? How could those images cohere in every direction so as to produce a true likeness of the object seen?[17]

And there are other, less familiar, phenomena that extramission can more credibly explain than an intromissionist account. Aristotle's theory of how rainbows are formed, for example, had relied crucially on his belief that ordinary atmospheric air, under certain circumstances, can actually reflect sight. In making this point, he had recourse to an unusual fact—complete with all the characteristics of what I will call in the next chapter a *trope*—which he narrated with a story:

> ἀνακλωμένη μὲν οὖν ἡ ὄψις ἀπὸ πάντων φαίνεται τῶν λείων, τούτων δ' ἐστὶν καὶ ἀὴρ καὶ ὕδωρ. γίγνεται δὲ ἀπὸ μὲν ἀέρος, ὅταν τύχῃ συνιστάμενος. διὰ δὲ τὴν τῆς ὄψεως ἀσθένειαν πολλάκις καὶ ἄνευ συστάσεως ποιεῖ ἀνάκλασιν, οἷόν ποτε συνέβαινέ τινι πάθος ἠρέμα καὶ οὐκ ὀξὺ βλέποντι· ἀεὶ γὰρ εἴδωλον ἐδόκει προηγεῖσθαι βαδίζοντι αὐτῷ, ἐξ ἐναντίας βλέπον πρὸς αὐτόν. τοῦτο δ' ἔπασχε διὰ τὸ τὴν ὄψιν ἀνακλᾶσθαι πρὸς αὐτόν· οὕτω γὰρ ἀσθενὴς ἦν καὶ λεπτὴ πάμπαν ὑπὸ τῆς ἀρρωστίας, ὥστ' ἔνοπτρον ἐγίγνετο καὶ ὁ πλησίον ἀήρ, καὶ οὐκ ἐδύνατο ἀπωθεῖν—ὡς ὁ πόρρω καὶ πυκνός· διόπερ αἵ τ' ἄκραι ἀνεσπασμέναι φαίνονται ἐν τῇ θαλάττῃ, καὶ μείζω τὰ μεγέθη πάντων, ὅταν εὖροι πνέωσι, καὶ τὰ ἐν ταῖς ἀχλύσιν, οἷον καὶ ἥλιος καὶ ἄστρα ἀνίσχοντα καὶ δύνοντα μᾶλλον ἢ μεσουρανοῦντα.[18]
>
> It seems that vision is reflected by all things that are smooth, and both air and water can be counted in this class. This happens in the case of air when it is condensed, but also often because of weak vision it causes a reflection without being condensed. This was the case with a man who suffered blurry and weak vision:[19] he always thought there was an

17. This last group of objections to intromission is conditioned by the fact that the main intromissionist theory available in antiquity was Epicurean, where what we perceive are films or images of atoms which peel off objects in every direction. See esp. Galen, *De plac. Hipp. et Plat.* VII.7.8f.

18. *Met.* 373a33f.

19. Later commentaries on this passage identify the man as a certain Antipheron (Olympiodorus says he is from Tarentum, Alexander that he is from Oreus). Aristotle elsewhere refers by name to a certain Antipheron of Oreus as being subject to a particular kind of madness that had him believing phantasms were real (*De mem.* 451a8), but Aristotle says nothing there about his eyesight, poor or otherwise. Michael of Ephesus, in the eleventh or twelfth century AD, explicitly tries to bridge the gap by pointing out that Aristotle's tale about the anonymous self-seeing man was really about the Antipheron of Oreus mentioned in the *De mem.* (*In parv. nat. comm.* 17.30).

> image before him as he walked, seeing a reflection in front of himself. This happened because his vision was reflected back to himself; thus the weakness and utter feebleness caused by his ailment made even the nearby air act as a mirror, and his vision could not pass through it—just as is the case with distant and dense air. This is why at sea distant landing-sites appear greater in size whenever the south-east wind blows, as also in mist, and so also the sun and stars look bigger when rising and setting than when high overhead.

That air will in itself be visible if it is sufficiently condensed is just taken as given here by Aristotle, as it will be later by Ptolemy, who compares it to the way that water, which is more condensed than ordinary atmospheric air, is itself visible rather than being completely transparent.[20] What is perhaps most interesting about Aristotle's use of the man who saw himself everywhere, though, is that it is very difficult to see how to make this phenomenon actually agree with the visual theory Aristotle himself proposed in other works.[21] On Aristotle's fully worked out (and particularly idiosyncratic) account, seeing does not in any way involve the extramission of visual rays, nor the intromission of images or light. It is instead the direct perception of the medium (air or water) by the eye, where the medium has itself been changed by the colors of the objects in the world such that it has the same form as they do. A green cup looks like a green cup because the form of its matter (green, cup-shaped) has altered the air between it and our eye in a green and cup-shaped way, and the air then presents these qualities to us directly. The self-seeing man has no place here.

In fact, any kind of reflection may provide difficulties for Aristotle's theory of vision, but certain characteristics of the case of the self-seeing man show that Aristotle had pretty clearly borrowed someone else's example to make his point in this instance. As Galen would later point out quite forcefully, the illusion Aristotle described makes sense only if one supposes—contrary to Aristotle's own visual theory—that the eye is sending out some force that, on encountering the resistance of the air, is then being bent back and trained on the subject himself, so that the man sees himself in the air before him, as in a mirror.[22]

20. Ptolemy, *Opt.* II.9.

21. Aristotle, *De sensu* and *De anima.*

22. Galen's objection is at *De plac. Hipp. et Plat.* VII.7.10f. One wonders here how someone like E. Cobham Brewer (author of *A Guide to the Scientific Knowledge of Things Familiar*, 1849) would have explained the physics of the so-called Brocken Spectre, which he saw as a reflection of the hiker in mountainous mist rather than as a shadow (see Brewer 1898, under "spectre").

I highlight these examples partly to show why extramission was so readily plausible in antiquity. Mathematical opticians do not always go to such lengths to justify their use of extramissionist theories, but it is nevertheless a historical fact that within the mathematical optical tradition, extramission of visual rays is taken for granted by pretty much everyone, and this is not just instrumentalist window-dressing. In Ptolemy's case, at least, it is underscored by a profound scientific realism.[23] Ptolemy, modifying the visual ray theories of Euclid and Hero of Alexandria before him, adopts a theory of vision which has a continuous visual "flux" emanating from the eye out to the world at large.[24] Sometimes Ptolemy represents this plenistic conical flux as reducible to discreet and linear visual rays for the sake of mathematical or hermeneutic convenience, but when push comes to shove, Ptolemy is deeply opposed to the discreet-ray model. His flux is a real physical plenum, though it is not possible on the available evidence to clearly associate it with any particular substance with any degree of certainty.[25] He treats the center of the eye as a point source of this radiation, and the resulting emission forms a conical shape as it emanates forward and outward from the eye, a cone whose base is out in the world at large, whose apex is in the eye, and whose angle at the vertex is exactly equivalent to the breadth of the visual field of the observer (fig. 5.1).

The Eyes as *Organs*

In order to understand what Galen is doing with the eye and its vision, it is first necessary to come to grips with an important theoretical superstructure that guides his own seeing in many instances. For Galen there is a very close link between the bodies of animals and their souls. It is not just that the soul is completely embodied and physical for Galen, but that the body can ultimately be defined as the proper *instrument* or *organ* of the soul (τὸ

23. Goldstein 1967; Musgrave 1981; Smith 1998; Jones 2005; Evans and Berggren 2006.

24. Euclid's rays are discreet, and objects can fall between individual rays when they are small or distant enough.

25. Smith 1988, 196, wants to see the flux as Stoic *pneuma,* the same substance Galen will have emanating from the eye, but this relies on a chain of inference which is not nearly so solid as I would like. That Ptolemy's *virtus regitiva* can be identified as equivalent to the Stoic *hegemonikon,* I am partial to accepting, but to move from there to the idea that Ptolemy's flux must therefore also be Stoic is to overburden the initial equivalence unduly. In discussing a dream I had, for example, I can easily use a phrase that is more or less equivalent to "my unconscious," without implying that I accept any other baggage of Freudian psychology (e.g., "my Oedipal complex").

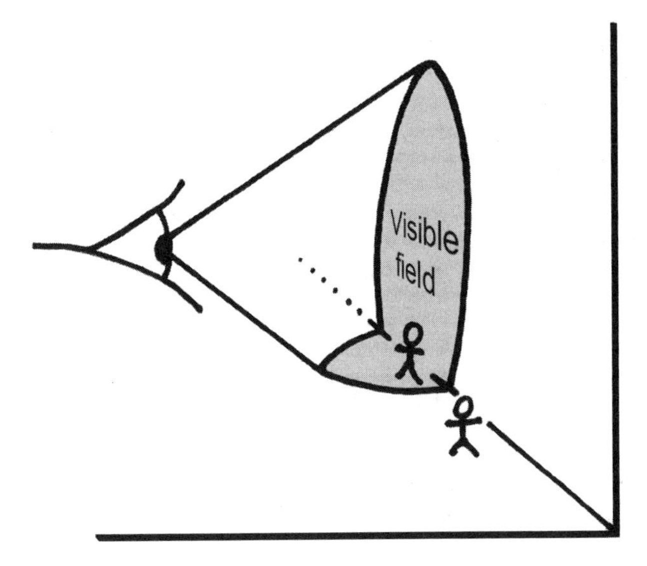

Fig. 5.1 The visual cone

γὰρ σῶμα ταύτης [*sc.* τῆς ψυχῆς] ὄργανον).[26] We can see this as the body's being the soul's tool for interacting with the world, and this in two directions: in terms of sensation and nutrition (internalizing) and in terms of communication and will (externalizing). Thus, says Galen, by means of their hands (which externalized their ideas as writing) and our eyes (which are internalizing their writings as ideas) we can even now converse with Plato, Aristotle, and Hippocrates.

To fully appreciate the thrust of the relationship between soul and body for Galen, we need to look a little closer at his use of the word I am translating as *instrument* or *organ*, ὄργανον. It is indeed the normal Greek work for tool or instrument of the hammer or file type, but in medical writing it is commonly used more specifically to refer to the parts of the body that perform functions of various sorts, and it is from this meaning that we get our English word for bodily organs.[27] As Galen defines it,

26. Galen offers a magisterial argument for the physicality of the soul in *Quod animi mores corporis temperamenta sequuntur*, even though he is there and elsewhere cautious about *genuinely* committing to the idea. See Menghi and Vegetti 1984; Hankinson 1991; von Staden 2000; Tieleman 2003.

27. ὄργανον is also the origin of the English *Organon*, which collectively refers to Aristotle's logical books (where logic is conceived of as a basic instrument for doing philosophy), and hence by extension also Bacon's *Novum organum*. On Galen's definition of the word, see May 1968, 67n3.

> ὄργανον δὲ ὀνομάζω μέρος ζῴου τελείας ἐνεργείας ἀπεργαστικόν, οἷον ὀφθαλμὸν ὄψεως καὶ γλῶτταν διαλέκτου καὶ σκέλη βαδίσεως.[28]
>
> I define *organ* as a part of an animal that is cause of a complete action, such as the eye for vision, the tongue for discourse, or the legs for walking.

Thus the body is the organ of the soul.

Where Galen's use of the word gets particularly interesting is when he pushes the breadth of meaning inherent in the Greek word to nuance our understanding of the soul's relationship to the body: In *De usu partium* I.2, Galen is discussing the various weapons and faculties that animals have for their protection, and he asserts that people, although they may look as though they are vulnerable, are actually the best protected of all. For they have the hand, which can provide sharper and stronger armaments than any animal is provided with by Nature:

> οὔκουν γυμνὸς, οὐδ' εὔτρωτος, οὐδ' ἄοπλος, οὐδ' ἀνυπόδετος ἄνθρωπος, ἀλλ' ἔστι μὲν αὐτῷ θώραξ σιδηροῦς, ὁπότε βούλοιτο, πάντων δερμάτων δυστρωτότερον ὄργανον, ἔστι δ' ὑποδημάτων παντοῖον εἶδος, ἔστι δ' ὅπλων, ἔστι δὲ καὶ σκεπασμάτων. οὐ γὰρ θώραξ μόνον, ἀλλὰ καὶ οἰκία καὶ τεῖχος καὶ πύργος ἀνθρώπου σκεπάσματα.
>
> But man is neither naked, nor vulnerable, nor without protection, nor barefoot, for he can have an iron *thorax* whenever he wants, an organ more protective than any skin, and he has every kind of sandal, of weapon, and of armor. And not only does man have the *thorax,* but he also has houses, walls, and towers as protection.

Here we see a deliberate juxtaposition of the artifices of Nature and the artifices of man. The skin, an organ of animals, is compared *qua* organ to the instruments of warfare: armor, weapons, walls, towers.[29] Not only is this comparison an insight into Galen's view of the state of nature as constant warfare between species and individuals, but it also shows how easily the concepts of bodily organ and man-made *technai* bleed into one another. This is further underscored by the fact that the improved organ he uses as an example straddles this same line between human artifact and the natural:

28. Galen, *Meth. med.* I.6.

29. David Langslow has reminded me here of how rich a source of metaphor warfare is in pathology, anatomy, and therapeutics.

thorax in Greek refers both to an armored breastplate and to the part of the human body the breastplate covers. Indeed, Aristophanes had earlier played on just this ambiguity in a passage in the *Wasps*.[30] The point here is not to take a McLuhian book-is-an-extension-of-the-eye position, but rather to underscore the fact that the bodily organs themselves are the products of the active and divine craftsmanship of capital-N Nature for Galen.

And this is where we can see one of the primary agendas that drives Galen's anatomical work in the *De usu partium* and elsewhere: by looking closely and intelligently at the structures and details of the body by means of theoretically informed and disciplined seeing we can see therein the wisdom of its creator. Nature—personified, active, wise—is credited with having created perfect instruments in animals.[31] Each part of the body shows the wisdom and the forethought of the creator, since each organ is so perfectly crafted for its function, and every organ is the best of all possible constitutions for the performance of that function.[32] This has two important implications: (1) Galen, in approaching his anatomical observations, will be looking through a frame or a filter of functional-teleological implications that needs to see how each detail contributes with maximal utility toward some end, and (2) Galen is frequently looking through his bodies at the creator: the physical is his window to the divine. But we need to be careful here: in making these two points, we are not saying that his vision is clouded, nor that the objects are transparent. Things are, as they should be, much more complicated than this.[33] To fully understand Galen's theory of how visual sensation comes into the eye, it is necessary to keep two related points in mind: the fact that the entire system was created by a wise, beneficent, personified Nature, and (partly derived from that) the bivalency inhering in Galen's use of the word *organ* itself.

For Galen, all sense perception and motor function relies in some man-

30. Aristophanes, *Wasps* 1190f.

31. On Galen's definition(s) of Nature, see esp. Kudlien 1981; Jouanna 2003; see also May 1968, 10f.; Hankinson 1989; and (albeit briefly) Nutton 2004, 234–35. Pigeaud 1993 is tangentially useful. On divine providence with respect to vision in particular, see *De usu partium* X; *De plac. Hipp. et Plat.* VII.5.13f.

32. By *constitution*, I want to underscore Galen's emphasis on the combination of the physical makeup of an organ, its particular balance of the elements, its geometrical shape, and the arrangement of its parts (compare Galen, *De usu partium* I.9).

33. On the relationships between Galen's teleology and his materialism, see Singer 1997; Hankinson 1989. In many ways, the questions raised by this issue are profoundly relevant to any reading of *De usu partium* in particular, where Galen is very explicitly running between a hands-on, detailed anatomical study, a teleological paean to a benign and wise creator, and, from time to time, a physical (elemental) explanation of the composition of organs and parts.

ner on a physical substance, pneuma, in the nerves of the body.[34] Seeing happens because the optic nerve conducts pneuma from the brain to the eye: "The brain sends part of itself out to the crystal-like humor [the lens of the eye] to know how it [the lens] is being affected."[35] This pneuma then passes through the various humors and structures of the eye itself to emerge from the pupil. But at this point Galen takes one of the standard ancient objections to extramission seriously: if vision is dependent on something that is physically sent out from the eye, how is it that it takes no time at all to see very distant objects like the stars, which we can see instantly upon opening our eyes?[36] Moreover, any pneuma extramitted would have to become impossibly dilute as it reached out to distant and large objects. In order to get around these problems, Galen proposes a unique theory: rather than the extramitted pneuma running off in all directions and to an infinite distance, it instead needs to reach only to the exterior surface of the eye, where it impacts the atmospheric air that surrounds it. It then *alters* the atmospheric air and uses that air as though it were an organ of the body:

> λείπεται οὖν ἔτι τὸν πέριξ ἀέρα τοιοῦτον ὄργανον ἡμῖν γίνεσθαι καθ' ὃν ὁρῶμεν χρόνον, ὁποῖον ἐν τῷ σώματι τὸ νεῦρον ὑπάρχει διὰ παντός.
>
> What remains, then, is that for the duration of the time when we are seeing, the surrounding air becomes for us just the sort of organ that the nerve always is in the body.[37]

A little later he elaborates: "The brain is to the nerve [lit. has the same ratio] as the eye to the air."[38] He compares the action of the pneuma to that of the sun, which touches the upper atmosphere and is thus able to illuminate the entire air completely and instantaneously (but note that this comparison is not an explanation of the role of light in vision—this Galen omits to offer us). There is a second respect in which the pneuma is sun-like (ἡλιοειδής): it is, Galen tells us, "luminous" (αὐγοειδής, lit. light-like) or "fiery" (πυροειδής, lit. fire-like). He offers empirical evidence for the luminosity of visual

34. The exact details of the precise role of *pneuma* are not always as clear as we might like, however. See Wilson 1959; Lloyd 1987, 213; von Staden 2000; Rocca 2003, 66.

35. ἀπέτεινεν οὖν ἑαυτοῦ τινα μοῖραν ὁ ἐγκέφαλος ἐπὶ τὸ κρυσταλλοειδὲς ὑγρὸν ἕνεκα τῆς γνώσεως τῶν κατ' αὐτὸ παθημάτων (Galen, *De usu partium* VIII.6).

36. He has something like this clearly in mind when he points out that alterations in atmospheric air, which is a continuum, should happen in an instant, ἐν ἀκαρεῖ χρόνῳ (Galen, *De plac. Hipp. et Plat.* VII.5.7).

37. Galen, *De plac. Hipp. et Plat.* VII.5.5.

38. Galen, *De plac. Hipp. et Plat.* VII.5.32.

pneuma by calling on the example of feline eyes. We are all familiar with the sometimes unsettling phenomenon of seeing a cat's eyes fairly glowing in the dark, when nothing else is visible. We now would dismiss this phenomenon as only an apparent glow due to reflected light, but we cannot do quite the same with this "observable fact" that Galen offers us: if you get, he says, a sufficiently large cat, say a lion or a leopard, and observe it in a darkened room, you can see not only its glowing eyes, but also a circle of light shining on its nose whenever the leopard looks that way.[39] Not only this, but the circle will be measurably smaller closer to the eyes and measurably larger toward the tip of the nose due to the expansion and contraction of the visual cone (he even offers us a mathematical law for relating the size of the circle to the distance from the eyes).

This example aside, Galen's account of vision is broadly speaking based on earlier Stoic (or more specifically Chrysippean) models,[40] although Galen importantly modifies the details of how perception itself occurs, criticizing the Stoic idea that the pneuma transmits information by encountering resistance, in the same way that a walking-stick might,[41] and instead turning to the luminosity (light-likeness) of visual pneuma as key.

Galen and Ptolemy agree that each sense will have one proper sensible that it perceives, although in the case of vision they differ slightly. Galen, like Aristotle before him,[42] thinks the proper sensible of sight is color, but Ptolemy qualifies this, arguing that although color is per se primarily and directly visible, it does not in itself provide all the visual cues we use. Indeed, for Ptolemy, seeing is seven-dimensional: we perceive not just color, but also "corporeity,"[43] size, shape, location, motion, and rest. These last six are perceived not by color alone (as Galen would have it) but more specifically by means of the boundaries between colors and any changes therein.[44]

Not Every Black Box Is a Camera Obscura

What happens in the eye itself to produce vision? Detailed as their own sides are, in both Ptolemy and Galen there comes a point in each of their

39. Galen, *De plac. Hipp. et Plat.* VII.4.18.

40. See *SVF* II.232f.

41. Galen, *De plac. Hipp. et plat.* VII.5.41, VII.7.20.

42. At *De anima* 418a11–14.

43. Borrowing the translation from Smith 1996, 71n1, who clarifies thus: "I have rendered *corpus* as 'corporeity' to indicate that it is not so much body as the fact that something is a body that is apprehended by sight."

44. Ptolemy, *Opt.* II.7

explanations where they defer to a common ancient causal trope, which can essentially be summarized as "like affects like." They invoke this explanation at just the point where we might want them to be considerably more specific about the details in the physical causal chain. It is not, however, that they are each looking for a quick and easy way of whitewashing their own ignorance of a (the?) crucial link in the causal chain that begins out in the world and ends in our conscious perception. We can see it instead as part of their shared intellectual heritage, their shared *world,* that they should be perfectly happy with invoking what they each think of as a universal physical law in order to secure the relationship between our perceptions and objects-at-large. Rather than accuse them of black-boxing a link in that relationship by pointing to a mechanism that is devoid of any clear empirical meaning, we should instead see them as invoking one of the most self-evident of ancient general laws as being, in fact, the most sensible mechanism for linking the two realms, and linking them in a way that is in principle error-free. Like-affects-like may still be a black box, insofar as neither Ptolemy nor Galen (nor anyone else, to my knowledge) works out the precise physical mechanics of its efficacy.[45] But it is a black box that sits in a blind spot, so to speak, and I am for the moment more interested in how the mechanism itself is used to bridge a gap in the causal chain than I am with the precise internal workings of the mechanism itself.

We saw earlier that for Aristotle, light is just the state of the air's transparency being active rather than potential. For Ptolemy, there is something else going on entirely, even if he does use Aristotelian language to explain it.[46] Instead of talking about transparency, Ptolemy has light and the visual flux, the *visus,* interacting to produce vision, and when he comes to explain exactly how, he says it is as though light were *form* to color, which is as matter: *ut forme coloribus quoque ut yle.*[47] But light is also itself physical, and it falls on the visible object (*cadit super eam*) in the same way that the visual flux does (*incidit super eam*), and the more of either, the better a thing is seen.[48]

In understanding this position, we are severely hampered by the loss of book 1 of the *Optics,* where (as Ptolemy tells us in book 2) he had explained

45. Compare Plato, *Timaeus* 45c.

46. Smith has argued that Ptolemy's analysis borrows considerably in particular details from Aristotle (Smith 1988, 200f.), but as in the case of Smith's claims of Stoic influence, I want to be more cautious.

47. Ptolemy, *Opt.* II.16

48. Ptolemy, *Opt.* II.18: *ex diversitatibus quidem quantitatis videtur res magis, quando plus incidit super eam claritas visus, aut quando lumen plus cadit super eam.*

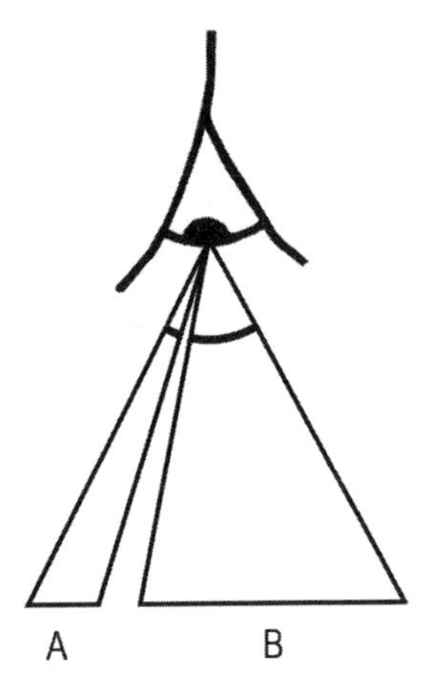

Fig. 5.2 Line B, being larger, subtends a larger angle than line A.

all of this in detail. There is also a nontrivial textual problem, in that the only text we have of the *Optics* is a Latin translation of a lost Arabic translation of a lost Greek original, with a possible Syriac intermediary between the Greek and the Arabic. Nevertheless, we can work out some of the particulars from the summary Ptolemy provides us, from his other works, and from how he handles particular puzzles in the rest of the *Optics*. Basically, light and color interact out in the world. The visual flux we send out is completely receptive ("passive" in an Aristotelian sense) and so cannot affect things in the world, but it can be affected by them because of something it has in common with light and color: it shares, to use the Aristotelian terminology of the Latin translation, the same *genus*.[49] But the visual flux itself is multidimensional. It perceives color primarily, to be sure, but it also perceives distance (apparently directly, through the length of the visual rays) and size (via visual angles). This perception of distance by length, and size by angle, is again framed in terms of an agreement of *genus*.[50] These different aspects of the visual flux are affected by the properties that are similar to—of the same *genus* as—them. Thus, for example, relative size is measured by the angle subtended by two objects, the larger subtending the greater angle when the objects are equidistant (see fig. 5.2).

The eyes (like all the sense organs) always tell the truth about how they are affected. They are not in and of themselves a source of error. How they were affected may provide false information about the world, as in optical illusions or the bending of vision in reflection and refraction, but the senses

49. *participat eis et in genere.* Ptolemy, *Opt.* II.23.
50. Ptolemy, *Opt.* II.63.

themselves do not lie.[51] The visual flux simply receives (*suscipiat*) the qualities in a manner that Ptolemy calls "straightforward," *simpliciter*.[52]

Galen gives us a slightly different fleshing out of the idea that properties and faculties must have certain specific similarities in order for veridical perception to occur. As in Ptolemy, something changes hands at the eyeball. Where for Ptolemy there is a stamping of *phantasia* somewhere in the eye by the passions of the visual flux, for Galen there is a passing of sensation from what we might call worldly or external pneuma to the internal, psychic pneuma that goes from the eyes to the brain. As for Ptolemy, the story is one of a continuous chain of causal linkages that hinges at one point on "likeness" as a mechanism in the transition of sensory impressions between the world and the body.

The basic idea is that psychic pneuma can be affected by worldly pneuma because they are the same kind of thing, and their affectations can thus pass from the one to the other in a simple and uncomplicated manner. We have already said that Galen's account of vision involves the interaction of psychic pneuma with worldly pneuma, and the use of that worldly pneuma as an organ of the body. So how, then, is the transition made between the worldly pneuma and our psychic pneuma? "In a word," says Galen, "like shakes hands with like," ἑνὶ δὲ λόγῳ τὸ ὅμοιον τῷ ὁμοίῳ γνώριμον,[53] or elsewhere, "like comes into communal effect with like," τῷ γὰρ ὁμοίῳ τὸ ὅμοιον εἰς τὴν τῶν παθημάτων ἀφικνεῖται κοινωνίαν.[54] The eyes perceive light because they are the "most light-like" of the organs (αὐγοειδέστατος). This "light-likeness" is the "luminosity" I referred to above as the fundamental property of visual pneuma that enables sight. Pneuma is said to be αὐγοειδής (light-like, luminous), πυροειδής (fire-like), and ἡλιοειδής (sun-like), each adjective here being a compound of the word εἶδος, *form* or *kind*. It is telling that Galen's descriptions of the powers of all of the senses are couched in terms of their sharing an εἶδος with their objects: touch is earth-like (γεώδης), hearing air-like (ἀεροειδής), and smell is vapor-like (ἀτμοειδής).[55] Even the very

51. See Ptolemy, *On the Criterion and Hegemonikon* (hereafter abbreviated as *Crit.*) 10.3, 11.1.

52. Ptolemy, *Opt.* II.23.

53. Galen, *De usu partium,* VIII.6. "Shakes hands with" is an admittedly imperfect rendering of a very difficult sense to translate. The adjective by itself means *well known, familiar.*

54. Galen, *De plac. Hipp. et Plat.* VII.6.10.

55. Galen, *De plac. Hipp. et Plat.* VII.5.42–44. Taste may appear to be a slight exception here, where its εἶδος has to do with receptivity to certain properties of its objects rather than to the objects themselves, so instead of being "food-like," it is said to be moist and "sponge-like" (σπογγοειδής). This is elaborated on a little later by Galen (VII.6.22) where he explains that the objects of taste are moist, from which we can see that the moistness of the organ of taste is again in a like-to-like correspondence with its objects.

relationship between the brain and the nerves is underscored by their sharing a form: Galen says that "the nerves are homo-εἶδος in substance with the brain," ὁμοειδὲς δὲ καὶ κατὰ τὴν οὐσίαν τὸ νεῦρον ὑπάρχον τῷ ἐγκεφάλῳ.[56]

"Likeness" thus stands as the fundamental mechanism underlying sense perception for Galen. More importantly, though, this likeness, or form-sharing, bridges the juncture between the two discrete systems, between internal and external, between psychic and worldly pneuma. So, too, for Ptolemy, it was form-sharing (*genus*-sharing) that bridged the gap between the colors that exist in the world at large and the visual flux being extruded by the eye, to allow those colors into our perceptive system.

Epistemologies of Seeing

The epistemological problems with vision and with the senses in general had long been pushed, both emphatically and persistently, by the Sceptics (Pyrrhonian Sceptics in particular).[57] Problems with knowledge and sensory experience stand at the very heart of Pyrrhonism. Sextus Empiricus, our best source for Pyrrhonism and likely a contemporary of both Galen and Ptolemy, takes the strong Sceptical position that physical objects are simply not apprehensible. In support of this, he offers a series of traditional arguments (called "ways" or "modes") that stand at the origins of Scepticism and that take clear aim at the senses.[58] Take eyes for example. People, dogs, and dragonflies all have different kinds of eyes, so surely colors are differently perceived by each species. How do we decide whose perception apprehends the object truly? Even within the same species individuals have different constitutions (humoral balances). One man is thin, another fat, so that things must appear different to the differently constituted bodies of one person versus another. One person gets pleasure or pain from very different things than another person does, which further shows that the same entities in the world affect different individuals differently. Does any of them

56. Galen, *De plac. Hipp. et Plat.* VII.5.13. For a compatible argument about Galen's caution in the face of medical empiricism, see Kuriyama 2002, 34.

57. Pyrrhonism is the stronger of the two ancient varieties of Scepticism, the other being Academic Scepticism (the school to which Cicero and Plutarch subscribed), which worried about the reliability of the senses, but did not take it as proven that they told us nothing about the world.

58. See, e.g., Sextus Empiricus, *PH* III.38f., I.35f. The word he uses for "apprehensible" is καταληπτόν, *graspable,* which is a technical term in Hellenistic epistemology often used to refer to perception, but whose meanings also included cognitive "grasping" (in a similar way to how the English "I see" also means "I understand"). The "modes," are collected and discussed most fully in Annas and Barnes 1985.

have access to the real object? Even our own senses disagree with each other about the nature of individual objects. Paintings feel flat but can look like they have depth. Perfume smells nice, but tastes awful. In any case, how do we know our five senses are even sufficient to be affected by all the properties of an object? (Comparing ourselves with dogs, whose noses are so very much better than ours, is telling here.) Our senses also seem differently affected when we are in different states: waking or sleeping, intoxicated or sober, healthy or sick. So, too, objects appear differently in different circumstances: columns look narrow at one end when seen from below, but symmetrical when viewed from the middle. A lamp looks bright at night but dim in the sunlight. Horn looks black on a goat, but its shavings are white. When are the senses grasping the reality of the object?

For Ptolemy, as for many of his contemporaries, part of the way around this was to reduce the problem of the senses to a problem of *judgment* (and once again, note the foregrounding of legal models). Just as a magistrate listens to the disclosure of evidence in order to decide the facts of a case by means of law and precedent, all with an aim to promoting social harmony, so the mind that investigates the world must use sense perception in order to adjudge what is (τὸ ὄν) by means of dialectic, all with an aim to finding the truth. Ptolemy maps the parallel as in table 5.1.[59]

Here Ptolemy's model is partly rooted in his fundamentally mathematical conception of science, and his account is saturated with mathematical terminology. He says that we can conceive of categories (1), (3), and (5) in this table as the terms (ὅροι) in a mathematical proportion, with (2) and (4) standing as the ratio or the (mathematical) means (διάστημα καὶ μεσότητες), which arrange things according to the relationships (διαδόσεις) between the extreme terms of the proportion (τὰ ἄκρα).[60] The nuances here are technical, but the basic idea is that sense perception mediates between *what is* and the mind in the same way that dialectic mediates between the

59. Long 1989 notes that some of Ptolemy's categories are shared by Sextus, of all people: specifically the middle three categories of instrument, agent, and means (although Sextus may be using category (4), means, with a slightly different emphasis than Ptolemy does). Long sees the two accounts as virtually identical, but Sextus' use of the application of sense impressions, ἡ προσβολή τῆς φαντασίας, in slot (4) where Ptolemy talks of dialectic as the means by which the intellect judges, is more clearly rooted in a particular set of debates centered on Stoicism. There are family resemblances between the two accounts, but it is difficult to see how exactly to map Sextus' version onto Ptolemy's.

60. Recall that Galen had also used a metaphor from proportion theory to compare the relationship between the brain and psychic *pneuma* to that between the eye and worldly *pneuma* (*De plac. Hipp. et Plat.* VII.5.32). Although similar in kind to Ptolemy's analogy, it differs insofar as it breaks down Ptolemy's categories 1–3 differently, inserting two intermediary objects between 1 and 3, and he does not reify their relationship in the same way that Ptolemy does.

TABLE 5.1

	Investigating a legal case	Investigating the world
1. Point at issue	Facts of case, ἡ πρᾶξις	What is, τὸ ὄν
2. Instrument	Disclosure of evidence	Sense perception
3. Agent	The magistrate	The mind, ὁ νοῦς
4. Means	Law	Dialectic, ὁ λόγος, διαλέγειν
5. Goal	Social harmony	Truth, ἡ ἀλήθεια

mind and the truth. But it also implies that the mediation is not the kind that filters, or represents, or even misrepresents, but is instead fixed, direct, and rigid. It brings the two terms together into a particular relationship, and is in fact the very substance of their interrelation.

And this interrelation is the essence of perception. Perception is not only the action of objects in the world on the senses; nor is it only the appearance of ideas in the mind. It instead arises as an interaction between those two systems: *per relationem et ratiocinationem inter illas factam et id quod dividitur et procedit a virtute regitiva,* where *relatio et ratiocinatio* may be best translated as "response and reckoning."[61]

What the mind receives, though, are not the actual changes in the sense organs, but something else, something called φαντασία, images.[62] Where the senses touch the world, Ptolemy tells us, the *phantasia* are impressions (lit. "pressings, molds," not "representations") which then pass sensation on to the mind.[63] But the picture he offers is one of continuous and unbroken contact, touching, from the mind out to the world at large. What the mind judges are not just the raw sensory impressions, though, but also the objects in the world that are the causes of those impressions—and it is here that both error and error correction can occur. Error, however, does not come

61. Ptolemy, *Opt.* II.22.

62. Ptolemy, *Crit.* 8.

63. Ptolemy, *Crit.* 2.4. On the Stoic understanding of *phantasia,* see Long 1999, 572–580. Long uses "impression" as his translation of *phantasia* produced by sensation, but does in other contexts offer "representation" as another translation, which could be seen as conflicting with my nonrepresentational reading of Ptolemy if it could be shown that Ptolemy is using *phantasia* in a strictly Stoic sense. Unfortunately, *phantasia* is a common term in Hellenistic accounts of perception, and Ptolemy's use of it is gray enough not to allow us to decide to what extent he may be committing to a Stoic view in particular here. Finally, I am not certain that *phantasia* as a blanket term in Stoicism for something like "the objects of thought" would be precluded from sometimes being representational (as in memory) and sometimes being nonrepresentational (as in perception).

about because the judging faculty is inherently faulty (Ptolemy tells us its *criterion* for judging is infallible),[64] but because it may be affected by impressions that for one reason or another are misleading. The senses always tell the truth about how they are affected, but those affections may still not accurately represent the true nature of the objects of those perceptions, or they may be of such a sort as to distract the judgment itself, tricking it into privileging inappropriate and misleading sources of information, as when it judges distance primarily (and inappropriately) by relative coloring or size and so is fooled by the kind of trompe l'oeil painting Sextus was worried about. This is essentially a problem of using one part of the visual faculty to judge aspects that are not of its proper genus: like using color or visual angle to judge distance.[65] Nevertheless, if we can understand what kinds of sense impressions create what kinds of images, and what kinds of objects create what kinds of sense impressions, the ruling part of the soul—called the *hegemonikon,* following Stoic terminology—will be able to work out, via the faculty of judgment, what the natures of objects really are, insofar as is possible.[66] It does this in two ways: (1) by comparing experiences from other senses (or from the same sense at different times or under different circumstances), and (2) by reflecting on the sensation through a process of reductive comparison (προσβιβάζειν) and uncovering (ἀνευρίσκειν). This process helps us not only to know what to do with conflicting or confusing perceptions when they arise, but also to recognize when the senses are reporting in a clear and uncorrupted manner so that we can simply accept their reports as "within the limits of human ability, most infallible," which reports the mind receives "with a miraculous, almost incredible power."[67] This, then, is the trick: learning to know when and how the senses can be fooled, and to know how they should be corrected or more carefully examined. This makes empiricism a learned, skilled practice, and at the same time the essential tool in the acquisition of true knowledge.[68]

Galen does not problematize the relationship between the mind and

64. Ptolemy, *Crit.* 9.6. For the centrality of the *criterion* in Hellenistic epistemology, and its relation to debates between dogmatic schools and Sceptics, see Long 1989.

65. Ptolemy, *Opt.* II.84f., 126f.; *Crit.* 11.

66. As Smith has remarked: "It would be fair to say that Ptolemy's ulterior purpose in the *Optics* as a whole is to correct such misjudgments by explaining them away" (Smith 1999, 40). There is a lovely ambiguity in this regard at Ptolemy, *Opt.* II.83: "We should draw distinctions between illusions . . . in order that we can solve doubts that come up in investigations *de scientia opticorum.*" Solve doubts that come up in investigations *concerning* the science of optics, or solve doubts that come up in investigations, *by means of* the science of optics?

67. Ptolemy, *Crit.* 10.5–6; *Opt.* II.73.

68. Compare Ptolemy, *Harm.* I.1, III.3.

sense perception to quite the same degree. Where for Ptolemy the mind judges *what is* by subjecting the evidence presented for it (sense impressions) to dialectic, Galen seems to be happy having the sense impressions carried by pneuma into the physical home of the *hegemonikon*, which he (like Ptolemy) locates in the brain. Physically running the chain of perception into where he knows the *hegemonikon* is located is, it seems, enough for Galen to be satisfied.[69] So long as we are not overly hasty in drawing conclusions from the senses, such as thinking we have seen Dion off in the distance when it was really Theon, there should be no problem.[70] But note that this is not so much a problem of hasty assent to *any* sensory impressions, as to the judgment of relatively complex phenomena as being true. It is not that we thought we saw Dion but we were hallucinating, but that the perception of the man we did see, Theon, was misattributed.

In Ptolemy's discussion, the problem of misattribution seems to happen at a more fundamental level. When he offers examples of misjudgments, they are not just that the man we saw was not the man we thought we saw, but more on the order that the man we thought we saw was actually a painting rather than a man at all. This is a deeper problem than Galen explicitly faces, and it cuts a little closer to the heart of the Sceptical assaults that both Ptolemy's and Galen's empiricisms were buttressing themselves against. It is that we could be misled not just as to details, but as to the true natures of the things in the world. Nevertheless, Galen could certainly respond that his no-hasty-judgments rule would still allow him to work out the difference between the painting and the man, just as Ptolemy recommends cross-referencing any sensory experience with other experiences to determine the truth.

The Centrality of Experience

For both Ptolemy and Galen, the problem is the same, and the answers overlapping. The overarching question is this: how do you justify a scientific methodology that roots itself so very deeply in empiricism, in the face of the radical Hellenistic Sceptical assaults on the reliability of sensory experience?

The answers for both of them rely on the development of tight causal

69. The precise correspondence between *pneuma* and consciousness is left open in Galen. Nevertheless, it is clear that consciousness depends on *pneuma*, and intelligence on the quality of the *pneuma* (*De usu partium* VII.13).

70. Galen, *De animi cuiuslibet peccatorum dignotione et curatione*, 6.

chains that run without interruption from the objects of the world into the mind itself. Those objects physically affect the senses, which in turn interact with internal physical mechanisms, which in turn physically affect the mind. The chain is unbroken and error-free, insofar as perceptual error is caused not by the senses misinforming us of how they have been affected, but by the mind misjudging what that affectation implies about the nature of the object that caused the sensation. Even this talk of "misinforming" and "implication" is in a sense misleading, insofar as it takes a particularly modern perspective for granted, the idea that in the first instance we have experiences that we take to represent the world in a certain way. Neither Ptolemy nor Galen is a representationalist in this sense.[71] Instead, they give us not so much sense impressions as sense reactions. Both Ptolemy and Galen can be seen in this light to espouse causal versions of what modern philosophers call *direct realism* with regard to perception: we perceive real objects in the world immediately and directly.[72] Ptolemy adds, in his discussion of the *criterion,* that we then make representational (language-rooted) judgments about *what is,* but for him the perceptions themselves are non-representational. But all of these perceptions, all of this seeing, is reliable only to the extent that it is subjected to careful rational control, and the more careful the better: veridical seeing is a rational skill.

In both Ptolemy and Galen, the chain of causation relies at critical junctures on the invocation of the law of like-affects-like as the primary mechanism for how the material of the world can be trusted to affect our internal states in nonrandom ways through the senses. But this is not to put phenomenological brackets around the relationship between the senses and the world in an attempt to avoid questions of what the exact nature of the relationship between the visual flux and color is, or the exact nature of the relationship between internal and external pneuma. It is instead to appeal to what both Ptolemy and Galen take to be a basic law about how the universe works: Like-affects-like looks to be a law of nature in Hellenistic philosophy, and one whose veracity neither Ptolemy nor Galen is worried about. It may look like a black box to us, but they clearly see it as an *explanans,* not as an *explanandum*-on-hold.

71. Ptolemy is explicit that judgment, insofar as it uses language, is representational, but sense perception itself is causal.

72. For accounts of this position, see, e.g., Strawson 1979; Putnam 1999; although why Putnam claims that a causal account of perception should be "wholly incompatible" with what he calls natural realism is opaque to me, unless it rests on his (unwarranted) assumption that causal theories are sense-data theories (12, 22)—here contrast Dretske 1995. In a different tradition, but making at least sympathetic points, see Gibson 1966.

Finally, we should notice the centrality of experience to both Ptolemy and Galen. Sense perception is for Ptolemy prior to thought in actualization: we experience before we can think. And animals, which are lower than us, can experience but not think.[73] True, the seat of the *hegemonikon* is the principle cause of living well, and so it is above the senses in that respect; but, Ptolemy tells us, the senses come a close second. And in the acquisition of true knowledge, although the mind is the primary agent, the senses are absolutely indispensable as *the* source of evidence.

Galen's corpus is littered with appeals to experience, and the two most common accusations he levels at the targets of his invective are that his opponents either cannot think straight, or that they have missed the empirical evidence that is before them. Experience is one of the fundamental cornerstones of knowledge: "I have thought it best to guard against nothing so carefully in argument as against positing something that contradicts what is clearly observed," ἠξίουν τε μηδὲν οὕτω φυλάττεσθαι κατὰ τὸν λόγον ὡς τὸ τοῖς ἐναργῶς φαινομένοις ἐναντία τίθεσθαι.[74] And the methodology Galen outlines for determining the location of the *hegemonikon* itself is what he calls the "demonstrative method," ἡ ἀποδεικτικὴ μέθοδος, translated (perhaps not unjustly) by Phillip de Lacy and by Teun Tieleman as "the scientific method."[75] For the details of this methodology Galen refers us to a treatise of his on the subject, unfortunately no longer extant, but we can work out the general approach by looking at how he applies this method in the solution to physiological problems. For example in attempting to determine where in the human body the governing part of the soul is physically located, his technique is highly (though not only) empirical: he looks for physical structures in the body that could be responsible for the kinds of activities the soul is supposed to perform (*viz.*, sensation and motion, for which he credits the nerves), and he then looks to see which organ is physically connected to the greatest number of these structures. "Everything," he tells us, "which falls outside this path is superfluous and invasive." ταύτης τῆς ὁδοῦ πᾶν ὅ τι περ ἂν ἔξω πίπτῃ, περιττόν τ' ἐστὶ καὶ ἀλλότριον.[76] Like Ptolemy, he sets the *hegemonikon* in the brain. And like Ptolemy, he sets vision up as the highest of the senses—though for very different reasons. For Galen it is because of the very great volume of pneuma used in vision (shown by the size of the nerves that lead to the eyes), whereas for Ptolemy it has to do with the physical lo-

73. Ptolemy, *Crit.* 9.2. We can also experience when we are not thinking.

74. Galen, *De plac. Hipp. et Plat.* II.1.1.

75. To be fair, it is not our scientific method, but it can be argued that it is *a* scientific method. For the translation, see De Lacy in Galen, *De plac. Hipp. et Plat.*; Tieleman 1992.

76. Galen, *De plac. Hipp. et Plat.* II.3.8.

cation of the eyes as the sense nearest to the brain. There is also the general *topos* they share with their contemporaries, as with many moderns, that unreflectively sees vision as the paradigm for the senses in general.

Whether all of this provides for either of them a bulletproof antidote to Scepticism is still an open question. What is clear is that both Ptolemy and Galen felt the need to work out in detail what they saw as a solid epistemological footing for experience in general, and for vision in particular, in order to ground each of their investigations. They both did this by running tight causal chains from the world into the brain. And Galen, at least, seems to have thought he had succeeded in diffusing Scepticism: in the autobiographical *Prognosis,* he recounts a memorable incident that happened to him while preparing to dissect a pig in front of an audience of philosophers. Galen launched into a description of the very delicate and careful anatomical demonstration in which he would lead the philosophers, explaining how he would show them that there was a very fine pair of nerves, one on either side of the larynx and each as thin as hairs, that were responsible for vocalization. If these were cut, it would leave the animal speechless. Here, to Galen's horror, a philosopher by the name of Alexander interrupted him:

> τοῦτο πρῶτον· ἄν σοι συγχωρηθείη τοῖς διὰ τῶν αἰσθήσεων φαινομένοις πιστεύειν ἡμᾶς δεῖν;
>
> But first, must we concede to you this point: that the evidence of the senses is trustworthy?[77]

Galen's response was to simply storm out of the room in a huff, his parting shot to say that he would not have come had he known that the ἀγροικοπυρρώνειοι were going to be there.

The insult he coined? A compound of *Pyrrhonians* and *boors.*[78]

77. Galen, *Prog.* 98.5.

78. The best translation of the word I have seen is by Gleason 2009: "scept-hicks."

CHAPTER SIX

The Trouble with Taxa

> Now, by the hidden and admirable Pow'r of the Loadstones, the Steel-Plates were put into motion, and consequently the Gates were slowly drawn; however, not always, but when the said Loadstone on the outside was removed, after which the Steel was freed from its Pow'r, the two Bunches of Garlick being at the same time put at some distance, because it deadens the Magnes and robs it of its attractive Virtue.
> —Rabelais 1653, V.37

We saw in the last chapter how the larger intellectual context in which Ptolemy and Galen were working allowed them to plug gaps in their explanatory chains for vision by using a law of nature, "like affects like," that the modern sciences would not be so happy to invoke. That neither of them saw what we see—how an important lacuna was hiding in the black box of like-affects-like—raises some interesting questions about how phenomena come to be problematized, both within and between discourse communities. In the present chapter, we will explore these questions more fully, with some rather unsettling consequences for the epistemology of experience.

Many of the natural phenomena discussed so far in this study have been fairly straightforward and familiar: optical illusions, volcanoes, earthquakes, birds. But as anyone who has spent a lot of time reading the premodern sciences knows, not all the phenomena discussed in ancient texts are so easily domesticable for the modern reader. A lot of strange things were thought to be happening—and happening regularly—in the ancient world.

A version of this chapter was published as Lehoux 2003a.

This poses some new problems of interpretation and understanding for us as readers of ancient texts. We are not only moved onto unfamiliar ground with some of the observations deployed in ancient sources, but we are faced with otherwise very intelligent, rational, and careful observers making incredulous or impossible claims. A standard example that one might parade here is Pliny's Monocoli, his one-legged, umbrella-footed people from the Himalayas.[1] We could try, if we wanted to, to come up with a "genealogy of error" in an effort to explain how this nonexistent race came to get a place in Pliny's *Natural History*, calling on exaggerated traveler's tales, the corruptibility of oral reporting,[2] or plain old gullibility.

I propose to do something different. I propose deliberately to avoid questions of truth and falsity, at least absolute or transhistorical ones. If the existence of Monocoli does not generally come into question in antiquity, then there must be some reason why they seemed so plausible, some reason why they fit with people's beliefs about the world. This is to some extent in line with more recent approaches to the history and sociology of science that look, for example, at how centers view their peripheries (so with the Monocoli, tracing increasing circles of oddness out from Rome itself, where races become, almost normatively, increasingly exotic with distance).[3] But not all the examples we will meet with are amenable to such a reading: some of them are rather bafflingly not peripheral at all.[4] How do we deal with cases closer to home, cases that should have been obvious to everybody? (To anticipate my conclusion: much hinges on the word *obvious*.) Taking a more bidirectional epistemological perspective, I propose to ask questions not just about why Pliny believed such silly things, but simultaneously why we think these things are silly.[5] Out of this analysis, an important symmetry emerges, one that sheds light not just on how the ancients knew the world, but also on how we know the world, and thus what we can say about

1. Also called *Sciapodes* or *Skiapodes* in other sources. See Pliny, *NH* VII.23.

2. Indeed, one attempt to isolate the source of the reports of umbrella-footed people has argued that Pliny had heard confused reports of people with elephantiasis (Routh and Bhowmik 1993). As one recent paper asks in frustration with this general approach: *is there no end to retrospective diagnosis?* (Karenberg and Moog 2003).

3. The literature on centers and peripheries is vast, spanning a number of disciplines from sociology to history, development studies, colonialism, and more. For our purposes here, we can cite (incompletely) Wallerstein 1974–89; Rowlands et al. 1987; Bilde 1993; Thomas 2000; Li Causi 2003; Barchiesi 2005; König and Whitmarsh 2007; Wallace-Hadrill 2008; Parker 2008.

4. As Beagon 2005, 43f. shows.

5. I see part of what I am doing in this chapter as a complement to Chang 2004, who is also interested in the question of how we justify scientific truths that are so close to our hearts as to be unquestionable, and what happens to our picture of the world when we examine them closely.

our knowledge of the ancients. To this end, I will in this chapter follow one implausible but ubiquitous belief about the world from its earliest attestation through to its rebuttal and ultimate oblivion. This trajectory will take us about as far outside the Roman period as we will go in this study, but there is good reason for it. The eventual complete disappearance of the belief turns out to be telling for many of the themes in this book.

The story itself begins with a meal: Plutarch (philosopher, biographer, priest) is having dinner with a few friends one evening in the late first or early second century AD. He and his upper-class circle get together like this regularly, and they eat, and they drink, and they talk. And they talk. Rich Greeks and powerful Romans, important visitors, friends, and eminent locals alike, they share a cultural, an intellectual, and perhaps most importantly a conversational background that includes familiarity with, and often explicit training in, the various schools of Hellenistic philosophy (including both the skills and the problem sets that come bundled with that training). Such people don't just have dinner, they have a symposium. In the middle of one of these intellectual feasts, Plutarch has the servants bring out a typically elaborate fish course to his guests, and one of them, Chaeremonianus of Tralles, points out a fish on the platter that looks like a remarkable creature he saw once while on a sea journey, a fish called an *echeneïs* (the Romans call it a *remora*).[6] That fish, it seems, has a noteworthy ability, which Chaeremonianus illustrates with a story. Once while he was sailing near Sicily, the boat he was on inexplicably slowed down all of a sudden. The reason for its sluggishness—discovered by a watchful sailor, who was seasoned enough to know what to look for—turned out to be one single remora sticking to the hull of the boat. When the sailor peeled it off, the boat immediately regained its full speed.[7]

In spite of the *echeneïs* just being true to its name (ἐχενηΐς in Greek means "ship-holder-backer," just as the Latin *remora* means "delay") some of Plutarch's guests find Chaeremonianus' tale risible. Nevertheless it does get them to talking about the physical force that causes things of this sort to happen: that is, the force known in antiquity as antipathy, the correlate and opposite of sympathy. Sympathy and antipathy (which almost always come as a package) are among the most ubiquitous of physical forces in our period, and they have a very wide explanatory and evidential sweep. It is sympathy that causes, for example, the strings on a lyre to vibrate by them-

6. *Quaest. conviv.* 641B.

7. On this power of the *echeneïs*, see Aristotle, *HA* 505b19, where it is reported that the *echeneïs* is also thought by some to bring success in love and law; also Pliny, *NH* 9.79.

selves when a corresponding note is sung nearby. Indeed, we still call this phenomenon sympathetic vibration, even if the sense of sympathy in our use is now only metaphorical. There were a host of other examples of observable phenomena that were said to be instances of sympathies or antipathies in antiquity, and Plutarch seizes on the story of the remora as an opportunity to list them. One example is amber. When rubbed with a cloth, it attracts threads and other small objects to itself (sympathy) unless those things have been wetted with oil, which is antipathetic to amber. We have no problem believing the observation underlying this example, for we know that rubbed amber attracts because of its static electrical charge (indeed, the Greek word for amber is *electron*). But the next commonly observable fact that Plutarch drops into the discussion is less tractable:

> ἡ δὲ σιδηρῖτις λίθος οὐκ ἄγει τὸν σίδηρον, ἂν σκόρδῳ χρισθῇ.[8]
>
> And the lodestone will not attract iron if it is rubbed with garlic.

Clearly, a modern explanation is going to be harder to come by in this instance, since the phenomenon itself is nonexistent. As we delve in to see why Plutarch believes it, however, we will raise questions about our own beliefs in its falsity, questions that prove to be contagious, infecting a wider range of our beliefs about the world than might at first be obvious. When it comes to knowing what is possible in the world and what is impossible, it turns out that the classification of entities does a lot more work than we might at first suppose.

Knowledge Claims and Context-Dependence

One important starting point is to underscore the extent to which taxonomic contexts matter. In the Plutarch story, the garlic-magnet fact emerges as an offshoot of the philosophical discussion of antipathy. If we look around for other instances of this strange claim (and it turns out that there are no small number of them in the millennium and a half after Plutarch), we see that when garlic and magnets come together as a pair, antipathy is never far away. Indeed, the garlic-magnet fact turns up almost exclusively in discussions of antipathy. Moreover, garlic and magnets come up almost always as corroborating empirical evidence for antipathy. This concomitancy is a dis-

8. Plutarch, *Quaest. Conv.* 641C5.

tinctive feature of the relationship between garlic-magnets and antipathy, which I will single out by calling garlic-magnets a *trope* of sympathy and antipathy, in the same way that, say, time travel is called a trope of science fiction. There are two aspects to this that are worth emphasizing: in science fiction, time travel is a standard tool in the literary repertoire, and also the use of time travel in a book or movie is generally sufficient (but not necessary) to categorize that book or movie as science fiction. A caveat: I do not want to suggest that tropes must be fictional or fantastical. They are simply deeply entwined in particular contexts. Sympathetic vibration, a perfectly factual and acceptable phenomenon, is just as much a trope of the sympathy-antipathy argument in antiquity as the garlic-magnet story.[9] Alternatively, to take modern examples: lateral gene transfer is a trope of molecular phylogenetics, and valence electrons are a trope of a certain theory of matter. Tropicality operates independently of truth and falsity.

Turning now to examples of this trope, we find that the next still-extant occurrence of garlic and magnets comes a couple of decades after Plutarch, when Ptolemy drops it into a discussion of antipathy in his great astrological work, the *Tetrabiblos*.

> ὥσπερ γὰρ τούτων ἑκάτερον ἐαθὲν μὲν δι᾽ ἀγνωσίαν τῶν ἀντιπαθησόντων, πάντη πάντως παρακολουθήσει τῇ τῆς πρώτης φύσεως δυνάμει, οὔτε δὲ τὸ ἕλκος τὴν νομὴν ἢ τὴν σῆψιν κατεργάσεται τῆς ἀντικειμένης θεραπείας τυχόν, οὔτε τὸν σίδηρον ἡ μαγνῆτις ἑλκύσει παρατριβέντος αὐτῇ σκορόδου.[10]

> Similarly, when each of these [sc. wounds, magnets] is left alone because of ignorance of its antipathies, it will inevitably develop according to the power of its original nature [sc. putrefying, attracting iron]; but neither will a wound undergo spreading or putrefaction if it is subject to the corresponding cure, nor a magnet attract iron after being rubbed with garlic.

As in Plutarch, Ptolemy invokes the example not for its own sake, not to alert an unknowing reader to the existence of the garlic-magnet phenomenon, but just the opposite. He uses garlic and magnets precisely because of its familiarity, to make a point about the similar workings of healing techniques. The point he is emphasizing is that an antipathetic substance has

9. On sympathetic vibration, see, e.g., Cicero, *De div.* II.33; Plotinus, IV.iv.41.
10. Ptolemy, *Tetr.* I.3.13

the power to prevent other causes from falling out as they otherwise would if left to their own devices. For Ptolemy, knowing this means that we can employ antipathy in our service; we have control over it. In this light, garlic-magnets are being employed to underline the efficacy of the force that also counteracts the natural degenerative progression of an ailment. Here healing is what we would call *allopathic,* specifically brought about through the judicious use of a physical force, antipathy. The trope of garlic-magnets is dropped into the discussion in order to call this particular class of phenomena, physical counteracting causes, most immediately to Ptolemy's reader's mind.

The next time this trope bubbles to the surface in the literary record is in the seventh century AD. An anonymous Greek alchemical treatise brings in garlic-magnets as an *explanandum*, for which the *explanans* is antipathy:

> οὐδὲν γὰρ ἀγνοεῖν χρὴ ὅτι κατὰ συμπάθειαν φυσικὴν ὁ μαγνήτης λίθος τὸν σίδηρον ἕλκει πρὸς ἑαυτὸν οὐδὲ ὅτι κατὰ ἀντιπάθειαν τὸ σκόροδον προστριβόμενον κατὰ τὸν μαγνήτην κωλύει αὐτὸν τῆς τοιαύτης φυσικῆς ἐνεργείας.[11]

> None should be ignorant that it is because of a natural sympathy that the magnetic stone attracts iron to itself, nor that because of antipathy garlic rubbed on the magnet impedes it in its natural action.

That magnets attract because of sympathy had long been, and would long continue to be, the standard explanation for their efficacy. That they can be impeded by garlic is brought in to complete the pairing of forces, since strongly sympathetic things are generally also strongly antipathetic with respect to other objects. The larger context of this passage shows that again, as in both Plutarch and Ptolemy, garlic-magnets are being invoked as a familiar example to fill out the range of the powers of the two forces. Sympathy and antipathy, the author is saying, are common—just look at all the examples.

A few centuries later, the great tenth-century miscellany known as the *Geoponica* includes a short treatise called *On Physical Sympathies and Antipathies,* which is attributed to "Zoroaster." It is precisely in a work with such a title that we might most reasonably expect to see our trope surface, and we are not disappointed:

11. Berthellot and Ruelle 1888, 2.428.21.

ἡ μαγνῆτις λίθος, ἤτοι σιδηρῖτις, ἐφέλκεται τὸν σίδηρον· ἐκπνεῖ δέ· σκορόδου προστριβέντος αὐτῇ· ἀναζῇ δὲ πάλιν τραγείου αἵματος ἐπιχυθέντος αὐτῇ.[12]

> The magnetic stone, or the lodestone, attracts iron, but it loses this power when garlic is rubbed on it. It returns to life again, however, when goat's blood is poured on it.

The addition of the remedy for the magnet is new here. But goat's blood as an active substance is another trope of the sympathy-antipathy argument. Not only this, but we can even work out why goat's blood should act this way on a magnet: look at Pliny's explanation of sympathy and antipathy:

> pax secum in his aut bellum naturae dicetur, odia amicitiaeque rerum surdarum ac sensu carentium . . . quod Graeci sympathiam et antipathiam apellavere, quibus cuncta constant, ignes aquis restinguentibus, aquas sole devorante, luna pariente, . . . ferrum ad se trahente magnete lapide et alio rursus abigente a sese, adamanta, rarum et opum gaudium, infragilem omni cetera vi et invictum, sanguine hircino rumpente.[13]

> Here the peace and war of Nature with itself will be told, the hatreds and friendships of things deaf and dumb, . . . which the Greeks call *sympathy* and *antipathy*, in which all things participate: water extinguishing fire, the sun evaporating water and the moon bringing it forth, . . . iron being drawn to the magnetic stone, and by another being repelled: adamant, a rare and delightful wonder, unbreakable and unconquerable by any other force, is smashed by goat's blood.

We see goat's blood being invoked here as antipathetic to adamant. But we know from book 37 of the *Natural History* that adamant works on magnets in exactly the same way that garlic does: robbing them of their power to attract.[14] Thus washing the magnet in goat's blood, a substance antipathetic to the kind of thing that robs magnets of their power, negates the original antipathetic power of the garlic, and so restores the magnets.[15]

The use of garlic-magnets in the ways we have seen so far, in lists of examples illustrative of antipathy, is telling in a couple of ways. On the one

12. *Geoponica* 15.1.28.2.

13. Pliny, *NH* XX.1–2.

14. Pliny, *NH* XXXVII.61; see also Albertus Magnus, *De mineralibus* II.i.1; Solinus, *De mirabilibus mundi* LII.56.

15. Della Porta 1589 works this out the same way (VII.54).

hand, it shows the authors assuming the phenomenon to be broadly familiar to their audiences. But calling it "familiar" is puzzling, since no such phenomenon exists. We might try "widely believed," but that would not capture the immediacy of the illustrative force that the examples look to have been calling on, if the uses we have seen mean anything. The puzzle deepens as we look more closely at the wider surroundings of our garlic-magnet claims, where other phenomena get paired with garlic-magnets, phenomena that are not just "widely believed" but in many instances are, as we should say, widely *known*.

Unproblematic Facticity

I have several times called garlic-magnets a *fact*. This is because, within its context, the trope behaves in exactly the same way as things we would be much more comfortable calling "facts." One indication in this direction is that we see the inclusion of garlic-magnets in enumerative arguments in exact parallel with facts of a more familiar sort (familiar to us at least). We have already seen one example in Plutarch where the garlic-magnet trope was paralleled with a true example: rubbed amber attracts objects except when those objects have been wetted with oil, and the magnet attracts iron except when rubbed with garlic. So also in Ptolemy garlic-magnets are paralleled in an enumerative argument with the fact that a wound will not spread or putrefy if it is subject to the corresponding antipathetic cure.

And we find that enumeration of parallel examples is a common way of discussing the explanatory range of sympathy and antipathy. As in Ptolemy and Plutarch, we continue to see parallelism in these otherwise unknown verses quoted in a scholium to Hesiod, again in the service of enumeration of effects:

> χρυσὸν γὰρ ἰὸς τῶν κυνῶν ἰοῖ μόνος
> ὡς τὸν μαγνῆτιν ἑλκτικοῦ θραύει σθένους
> ἡ σκορδικὴ δύσπνοια προστετριμμένη.[16]

> Only the spittle of dogs will tarnish gold,
> just as the stench[17] of rubbed garlic
> robs the magnet of its power to attract.

16. Scholium to *Op.* 109.

17. A second scholiast has commented on this word by explaining that δύσπνοια, *stench*, implies δύναμις, *power.*

Here the word for spittle or venom, ἰός, is a pun on "tarnish" in Greek. That spittle can tarnish gold if the dog is mad is confirmed also by Sextus Julius Africanus at the beginning of the third century AD, who cites the phenomenon as a good example of antipathy.[18]

Another example of parallel enumeration is doubly striking because it comes in the middle of what has often been taken to be a Sceptical argument. This one comes from Cicero's *De divinatione.* Cicero says:

> ut enim iam sit aliqua in natura rerum cognatio[19]—quam esse concedo; multa enim Stoici colligunt; nam et musculorum iecuscula bruma dicuntur augeri, et puleium aridum florescere brumali ipso die, et inflatas rumpi vesiculas, et semina malorum quae in iis mediis inclusa sint in contrarias partis se vertere, iam nervos in fidibus aliis pulsis resonare alios, ostreisque et conchyliis omnibus contingere, ut cum luna pariter crescant pariterque decrescant, arboresque ut hiemali tempore cum luna simul senescente, quia tum exsiccatae sint, tempestive caedi putentur.[20]

> Yet I do concede that there exists some kind of sympathy in the nature of things. And the Stoics have drawn this inference from many [examples]. The livers of mice are said to grow larger in winter; and dried pennyroyal to bloom on the exact day of the winter solstice, and its inflated seedpods to burst and the fruit seeds that are contained in them to spread themselves out in all directions; and strings in a lyre to resonate when different ones are struck; that it befalls oysters and all shellfish to grow with, and decrease with the moon; and trees are supposed to be best felled in winter when the moon is waning, for then they are dry.

We should resist the temptation to think that the use of parallelism in this instance was an attempt to bolster the questionable facticity of some wild claim by placing it alongside more pedestrian—and so epistemologically less problematic—facts. That this is not the case becomes clear when we compare how different facts are enumerated in our different authors (table 6.1). What we see is that each author sets beliefs we would see as true up against beliefs we would see as fantastic, but with no indication of episte-

18. Sex. Julius Africanus, *Cesti* 9.1.15, quoted in Michael Psellus, *Opuscula logica, physica, allegorica, alia* 32.27.

19. Following the MSS., Ax 1938 emends it to *contagio.*

20. Cicero, *De Div.* II.xiv.33.

TABLE 6.1. Parallelism in enumerative arguments for sympathy and antipathy

	Pliny	Cicero	Ptolemy	Plutarch
$Fact_1$	Water extinguishes fire	Mouse livers grow in winter	Wounds won't spread if treated with proper cure	Sight of a ram stops a mad elephant
$Fact_2$	Sun evaporates water	Dried fleabane blooms at winter solstice	Magnets will not attract if rubbed with garlic	Oak twig stops a viper
$Fact_3$	Magnets attract iron	Strings of lyre vibrate when others struck		Wild bull quieted if bound to a fig tree
$Fact_4$	Goat's blood smashes adamant	Oysters grow and decrease with moon		Rubbed amber attracts, unless objects oiled
$Fact_5$		Trees driest in winter under waning moon		Magnets attract iron, unless rubbed with garlic

mological distinction between them. This is no mere rhetorical trick on the part of our authors. Rather, the way in which the formula of listing is employed in each example indicates that the authors and their intended audiences see no epistemological difference between the various items in the list. Indeed, the arguments of which they are a part only work on the assumption that all the items in the list are equally true. Even when the particular causal line comes under fire, as it does with the remora in Plutarch, it is not the examples that are questioned, but the line of causation. The fact of the remora attached to the slowed ship is not a point of contention for the sceptical Plutarch, just the inference that the fish itself is the cause rather than an encrustation of seaweed that both slows the boat and attracts the fish.

Problems with Experience

We can now move back a step and ask why it is that the ancients think that garlic interferes with magnetism. Perhaps the most obvious answer to suggest itself here builds on an idea similar to the theory-ladenness of observation, only in this case without the actual *observation* part: some theoretical considerations about the nature or powers of magnetism and the nature or powers of garlic (perhaps attraction and repulsion, respectively)[21] led the ancients to posit that each of these two substances should cancel the powers of the other. Opposites negate. Then a fact is generated from theory, and the theory in turn further supported by that fact. Or so it would seem. But we need to keep in mind that this epistemological story depends on our standing in a particular relation to the phenomenon ourselves, such that the reason garlic-magnets need explaining—or at least explaining in this way—is that *we* know the phenomenon to be false. The privilege of standpoint is precisely why our explanation does not agree with how the ancients tell us they came by this belief, which in the end comes as a bit of a surprise. Look back at the passage from Plutarch that started us off:

> ἡ δὲ σιδηρῖτις λίθος οὐκ ἄγει τὸν σίδηρον, ἂν σκόρδῳ χρισθῇ.
>
> And the lodestone will not attract iron if it is rubbed with garlic.

In the very next sentence he tells us how he knows this fact to be true:

> τούτων γὰρ ἐμφανῆ τὴν πεῖραν ἐχόντων.
>
> We have palpable experience of these things.

In the Greek, the word I have translated as "palpable" in this passage means "visible, manifest, physically present for all to see." In legal contexts, it is used of a witness brought bodily before a court, or of evidence physically held up before the eyes of the jury. The word for "experience," πεῖρα ("test," "trial," even "experiment"),[22] is being used idiomatically with the verb "to have" just as it is in English: once we have seen something, we have had that experience. Plutarch is making a very strong claim to visuality. This fact is something that he says is known with the eyes. Not just testable

21. Bartholomaeus Anglicus (1495) certainly seems to think of garlic in these terms: *Item allium virtutem habet aperiendi, diuiendi, & incidendi humoros grossos & consumendi* (*De proprietatibus rerum* XVII.2), "And Garlicke hath vertue to open, and to tempre, and to dyvyde and to departe, kytte and waste grete humours and thyck" (in de Worde's translation).

22. It is a common assumption outside of scholarship on ancient science that there are no experiments before the Scientific Revolution, but this is increasingly seen by specialists in ancient science as misguided. One need only look at Galen's proof that the arteries contain blood, or at the sophisticated experimental technique of Ptolemy's *Harmonics.* See Barker 2000.

and repeatable under strained or exceptional circumstances, but seen all the time. That claim is a little surprising.

We can if we like try to reclaim our vantage point for explaining how Plutarch has gone wrong, but the *explanandum* has now shifted a little and we are consequently more limited in our available explanations. One option is to suppose that Plutarch is deliberately misleading his reader, but this would run in the face of the other examples of garlic-magnets as well as all the other facts like it that we have seen in the lists above, and consequently lead us into a culture (or a conspiracy) of delusion among ancient authors. Unless there is a good indication otherwise, it seems better to trust that our authors really believe what they report as true and that they think they have good reasons for believing these things, just as we trust our own readers to do the same for us. Alternatively, we could try a revised version of the theory-generation of facts, but with an added concomitant about how deeply the theory has ingrained this belief. After all, it is not that Plutarch just thinks the phenomenon is possible, but that he thinks that he (or at least someone close by) has experienced—actually seen—garlic destroy the attractive power of a lodestone.[23] While this overconfidence hypothesis rings truest to the modern ear, it is a reminder of just how deeply and invisibly the category of "experience" is intertwined with the category of "theory." But it is not just Plutarch who reifies theory as experience in this way, as we shall see.

Plutarch's appeal to experience for garlic-magnets is not unique. A similar claim, and one even more strongly worded, shows up fifteen centuries later, just before the belief in garlic-magnet antipathy disappears from serious scientific discussion forever. Alessandro Vicentini in AD 1634 composes an argument against occult qualities in which he says that garlic negates the power of a magnet, and that this is known by experiment.[24] Two years later, Bernardo Cesi says the same thing: *retundi vires magnetis allio, experimentis discimus quotidianis,*[25] "We know by daily experiments that the power of the lodestone is weakened by garlic." Albertus Magnus, in his *De mineralibus,* had earlier also claimed that the diamond-magnet antipathy is known by experiment, and Albertus' reported tests get reiterated and emphasized

23. An anonymous referee correctly pointed out that there is logically another possibility: that Plutarch may mean something significantly different by "experience" than we do. But Plutarch's wording, his use of πεῖρα in the Greek leans toward the implication of an actual *test* of the theory rather than just an observation made in passing. On ancient empirical arguments generally, see Lloyd 1979, chap. 3.

24. Vicentini 1634; Thorndike 1923–1958, 3:310–11.

25. Cesi 1636, 40.

(*clarissimis experimentis* . . .) by Cesi. Finally, as late as 1653, Arnold de Boate (an antischolastic, by the way) is still propounding the garlic-magnet antipathy in a work on minerals:

> Though the Adamant be the hardest of all stones, yet it is softned with Goats bloud, and there is a special Antipathy between that and the Load-stone, which . . . hath an admirable vertue not onely to draw Iron to it self, but also to make any iron upon which it is rubbed to draw iron also. It is written notwithstanding, that being rubbed with the juyce of Gar-lick, it loseth that vertue, and cannot then draw iron, as likewise if a Diamond be layed close unto it.[26]

De Boate's phrasing, "It is written," alerts us to another factor in the longevity of the garlic-magnet fact: these early modern authors have, on the face of it, an apparently more complicated set of reasons for accepting garlic-magnet antipathy than did Plutarch. They may be accepting the fact not just because of its plausibility under a certain worldview, but also perhaps because of the authority of what has by their day become a long-standing textual tradition. But even if this is the case, Albertus, Cesi, and Vicentini still do make the claim to empirical justification. Why? It would seem that, like Plutarch, they are allowing other categories, in this case both inference and testimony, to bleed over into the category of experience.[27]

Cesi and Vicentini's claims for experimental proof of the theory stand out all the more sharply, coming as they do half a century after della Porta is supposed to have disproved the effect of garlic on magnets—by experiment, no less—in his *Magiae naturalis* of 1589.[28] Della Porta says:

> Sed quùm haec omnia experirer, falsa repperi; nam non solũ flatus, & ructus alliorum magnetem à suo trahende munere non distrahebant, sed

26. De Boate 1653; de Boate was writing pseudonymously as "D. B. Gent.," in Plat 1653, 218.

27. On the epistemological questions surrounding testimony, see chapter 4, above. See also Quine and Ullian 1970; Coady 1992; Kusch and Lipton 2002.

28. To be sure, Cardano had earlier raised serious doubts, saying in essence that the purported action of garlic on magnets is false, "except maybe a little bit in the case of weak [lodestones]": "Nec, ut fabulantur, allio, caepisúe impeditur, multò minus adamante: nisi forsan adeò leuiter, ut in minimis solùm ac debilibus depraehendetur, in reliquis autê sensum effugiat" (Cardano 1582, book VII). Note also that della Porta's appeal to experience is still not what Dear 1995 would call an event experiment: although it is a specific claim to a particular experience, it is more like Chaeremonianus's report about the *echeneïs* than it is like Pascal's about the barometer. And della Porta's account does not serve as a touchstone or a proof for future commentators. Instead they all just return to the old and familiar kind of general experience claim.

> totus allij succo peruunctus, ita priores functiones obibat, ac si nunc; allio perungeretur, ut ferè nulla.[29]

> But when I tried all these things, I found them to be false: for not onely breathing and belching upon the Loadstone after eating of Garlic, did not stop its vertues: but when it was all anoynted over with the juice of Garlic, it did perform its office as well as if it had never been touched with it.

Likewise, William Gilbert begins his monumental *De magnete* (1600) by mocking the garlic-magnet story openly. Though he does not claim to have tested it himself, he does say (later in chapter 1) that it is by his experiments that the nature of the lodestone will be revealed, and thus will such silliness be disproved. Once experiment has shown us the *physiologia,* the folly becomes patent.[30]

In their counterarguments, both della Porta and Gilbert notably divorce garlic-magnets from their traditional taxonomic context. For della Porta and Gilbert, garlic-magnets fail to find a footing in their new ontological and classificatory setting. Antipathy is no longer the phenomenon under investigation, but instead a new, pared-down entity is: the power of the lodestone. The garlic-magnet story had to try to account for itself in this new territory, and it failed. A recontextualization of the lodestone had already begun a few centuries earlier: as early as the late thirteenth century, just a generation after Aquinas, magnets were beginning to divorce themselves from sympathy-antipathy: Pietro d'Abano cites the magnet as an instance of *attraction,* rather than of sympathy directly.[31] Jacques Rohault's mid-seventeenth-century rejection of the effect of garlic on the magnet, still being published in English as late as 1735, may at first look like a throwback to the old tropic context, insofar as he talks about "what some have called the *Sympathy* and *Antipathy* betwixt the Load-stone and Iron,"[32] but this deliberate archaism comes late in a chapter in which he has already been primarily discussing the "powers of the lodestone." Sir Thomas Browne moves the garlic-magnet antipathy onto even more difficult turf. He neuters

29. Della Porta 1658, VII.48, English trans. of Young and Speed 1658.

30. Or, to be more precise, once Gilbert's testimony of the experiments has convinced us of a particular *physiologia,* then that theoretical understanding will allow us to infer that the story must be false.

31. D'Abano 1476, Differentia li.

32. Rohault 1735, II.iii.32, emphasis in the original.

it by contextualizing it as simply "common public folly" in his *Pseudodoxia epidemica* of 1646:[33]

> But certainly false it is what is commonly affirmed and beleeved, that Garlick doth hinder the attraction of the Loadstone; which is notwithstanding delivered by grave and worthy Writers; by Pliny, Solinus, Ptolomy, Plutarch, Albertus, Mathiolus, Rueus, Langius, and many more. . . . But that it is evidently false, many experiments declare.[34]

Notice again the appeal to experiments.

Twenty-five years after Browne, Rohault also makes an explicit, if hyperbolic, claim to experiment: "Quant à ce que quelques Ecrivains rapportent que l'ayman n'attire pas le fer à la presence du diamant, ou que l'oignon & l'ail luy font perdre sa vertu, ce sont des contes, qui sont démentis par mille experiences que j'ai faites."[35] These things are refuted "by a thousand experiments that I have performed." Note how Rohault's wording mirrors Cesi's earlier phrase, "we know by daily experiments"—but of course Cesi's experiments were supposed to prove exactly the opposite fact to Rohault's. Della Porta likewise claims to have disproved the theory by actually trying it out, and by doing so in several combinations and permutations (anointing, breathing, belching). But then again Plutarch had made a strong appeal to experience when he asserted the phenomenon in the first place.

The Lab Section of the Chapter

Who do we believe and why? The answer is so very clear that few of us will feel the need to run off and find lodestones to bring into our kitchens for the big experiment. Those of us who do so will likely feel a little silly in the doing—imagine being caught beside the office coffee machine with a clove of garlic in one hand and a sticky, smelly magnet in the other. Picture the look on your bemused colleague's face as your explanation unfolds. What none of us imagines when we run this scenario in our heads, however, is the

33. Note the contagion metaphor in the title.

34. Browne 1646, II.iii. Browne's list is hasty and it looks as though he is simply naming all the authors he can think of who talk about the properties of garlic or magnets at any length at all. Not all of these authors mention the phenomenon of garlic destroying magnets specifically, and he misses some important ones who do, e.g., Nicholas of Cusa (*Idiota de sapientia* I.16). Matthiolus 1554, book V.cv, does note the phenomenon in his discussion of the magnet, adding that rust also deadens the magnet's virtue.

35. Rohault 1671, III.lxx.

colleague's bemusement turning to amazement when the magnet suddenly ceases to work. Even the eccentrics among us who give it a try, even they, know exactly how they expect the experiment to end.

But still we should remember that—even for the eccentric empiricist—the test only becomes necessary under the artificial conditions I have created in this chapter.[36] We know the falsity of garlic-magnets so immediately that no test was necessary at all until the problem deepened as the chapter developed. And even then, most of us still don't see the need. We know exactly where the disproof lies—in experience—and we know that so powerfully as to simply leave it at that.

The proof that it is false is empirical. It may be a strange kind of empirical argument that never needs to come to the lab, but it is still empirical for all that. On careful analysis we can argue that this empiricism is indirect, and indirect in two ways: one is its easy and unproblematic deferability, and one is its working by proxy. Every experience we have ever had with magnets, and every experience with garlic, leads us to know that putting the two together and expecting significant results would be some kind of category mistake. Inference from experience thus easily bleeds over into experience itself. Our experiences of magnets, and our experiences of garlic, are quietly but very firmly mediated by our understanding of magnets and our understanding of garlic, just as Plutarch's experiences of those things were mediated by his own understandings. But this is exactly where we hit the big epistemological snag: our argument against the garlic-magnet antipathy is no stronger, and more importantly no more or less empirical, than Plutarch's argument for it.

Have we ever been modern?[37] It is often said that our epistemology changed radically in or around the seventeenth century, but in this instance, it seems all we did was to shuffle classifications.[38] Indeed, if it were otherwise, we should have to think of a book like della Porta's colorful *Magiae*

36. In presenting this argument to different audiences over the years, it seems that some people will, after being provoked like this, make a point of actually running the experiment (the Geology Department's display case at Dalhousie University still smells faintly of garlic after one of my enterprising students decided to be a purist and not settle for modern refrigerator magnets). From time to time, someone invariably comes back to me and reports that it doesn't work, and I have no reports to the contrary. Now I have testimony to add to obviousness, but even if I did have a contrary report, I could question the reliability of the individual witness.

37. Or, as Hacking might put it: did our science ever mature? (Hacking 1979).

38. On the change in the ontology of kinds, see Hacking 1993. Kuhn, in his later work (Kuhn 1987, 1993), was beginning to move to an idea of taxonomic classifications as a way of dealing with the sticky problem of incommensurability. Taxonomy is at best, however, an analogy for how classification is working in the present instance.

naturalis as one giant "to-do list" for empiricists. Della Porta may claim to have put garlic-magnets to the test, but he certainly didn't apply such a scepticism broadly. His book is a virtual A-to-Z of things no rational person today could possibly believe but which have, for all that, never been tested. Did anyone ever empirically disprove the claim that there exists a certain unidentified kind of bird, "like a blackbird," which is generated by the putrefaction of sage? Instead, when the class of things that were generated spontaneously came to be seen as null, this bird quietly moved to the realm of fiction, and it did so along with the geese, frogs, snakes, mice, bees, eels, and all the other members of the set of things generated by putrefaction. So also the claim that bears love honey because it improves their eyesight: the bees that sting the bear's mouth serve to draw down the thick humor that otherwise clouds the poor bear's normally bad vision.[39] Certainly no empirical test, no "crucial experiment" involving bears and bees, can be cited (and here, I fear, even our eccentric empiricist will be at a loss). As with garlic-magnets, the implausibility of the bear's improved vision stems from the fact that we no longer think of ailments as manifestations of humoral imbalances, nor of stings and infections as moving those humors about. But under differing sets of classifications of the causes of disease, such things just do or do not make perfect sense. One could go on and on here, but the point is by now abundantly clear. Plausibility and obviousness, not to mention experience itself, have much to do with classification.

The Question of Worlds

At both ends of its life history, the garlic-magnet antipathy is sandwiched by claims that it is empirically justified. We are equally certain, and for exactly the same reasons, that it is not.[40] The claims to empirical justification on both sides are not disingenuous, but they are and should be worrying. We

39. Compare Pliny, *NH* VIII.129.

40. Jay Foster illustrates one way of reading this in his comments on a draft of this chapter: "We have the modern who says 'I don't need to rub garlic on a magnet, because I know it will not do anything' and the pre-modern who says 'I don't need to rub garlic on a magnet, because I know it will wreck the magnet.' They are offering exactly symmetrical claims. Neither feels that detailed justification is required, because the proposition is self evident. Justification is taken to mean: (a) rational coherence, (b) empirical testability, and if we accept the recent philosophical turn, (c) testimony. Both sides claim justification by (b) experience, but they really mean that they are offering a justification by (a) rational coherence. In this case, (b) experience is being evacuated into (a) rational coherence. The remaining interesting feature is that modern and pre-modern claims are so wildly different that it is bewildering how a criteria like 'rational coherence' could be applied. That is, they are just such different claims that it throws the conditions and criteria for 'rational coherence' up in the air."

may not have had the specific experience that would prove what we want to prove, but it is so obvious that we may as well have. Moreover, it is supported by a host of other experiences that we have actually had (or at least we think we have). Is such a foundation empirical? Strictly speaking, no. But the problem is that its obviousness renders its untestedness invisible, for both sides, and even when we manage to determine that this blind spot *is* a blind spot, we discover the problem to be so widely ubiquitous as to render exhaustive testing impossible. Think, too, about how we explain Newtonian mechanics by alluding to the behavior of billiard balls: how much do we really know about billiard balls? How far is the nearest billiards table? And whose playing are we using to serve as any kind of standard for demonstrating the laws of physics? Empiricisms that are not as empirical as we think are rather thick on the ground. I am not saying that this is something to fix. Just the opposite, I am arguing that it is both unavoidable and ubiquitous. But it is important to be aware of the fact that theory does not just underlie and inform observation; worldviews blind us—inevitably—to our sloppiness with the very category of the empirical.[41]

When we floated potential reasons for Plutarch's inclusion of garlic-magnets as a fact, we limited the list to explanations that presupposed the correctness of our own standpoint. But now the symmetrical messiness or incompleteness of empiricism appears to block such a move. We can reestablish some small degree of security by running off and getting our hands dirty with garlic, but the victory is limited to just the one small question. Bears and bees and goats' blood—questions as impractical as they are unfundable—will continue to loom, and if not them, then something else. So what happens when the standpoint becomes undermined in this way? Are we doomed to a run into relativism?

For now we might employ, in a very loose sense, some Hackingesque or Kuhnian talk of "worlds" (a theme I develop most fully in chapter 10). It would be very satisfying to be able to say something like this: because of the depth and invisibility of classifications, and because of their long reach into ontology and possibility, people with significantly different classifications can best be said to inhabit different worlds. Plutarch lives in a different world than we do, one with different entities, different rules, and different possibilities. A cautious reading might try to parse the difference of worlds here as only a change in ontology, so that "the power of the lodestone" (call it "magnetism") was a newly existent thing (force) that inhered in magnets but not garlic and thus served to distance garlic from magnets. But this

41. See also Quine 1953; Goodman 1978.

would be to tell only part of the story. The answer to the problem is not just located in ontology, for the force of magnetism, replacing sympathy, served as a kind-predicate. Magnets used to be the kind of thing that was sympathetic, as was garlic. Now magnets are the kind of thing that are magnetic, and garlic in our experience is not. While we do see a shift in ontology here, the real work is not being done by the new force *qua* entity, but by the new force *qua* class of thing to which magnets belong. Garlic and magnets had been of a kind in Plutarch's day, and during the scientific revolution the lodestone got reclassified.

To put it in a philosophical idiom: it is as though physics were predicating two objects as green, and then switched sometime around the sixteenth century to parsing one as "grue" and the other as "bleen."[42] Where garlic and magnets had been both green, they are now one grue, one bleen. They no longer go together in the same suit of clothes. But reading it in these terms spins us right back to Plutarch's different world: "to project 'grue' and 'bleen' rather than 'green' and 'blue'—would be to live in a different world."[43] Different classes, different possibilities, different entities. What is remarkable, though, is that both before and after this change occurred, green-blue on the one hand and grue-bleen on the other were thought to be tied to experience in exactly the same way and, curiously, to experiences that no one making or accepting the claims seems actually to have had. Classification was doing the epistemological work, but experience was getting the credit.

Epilogue

The kind of experience to which both the pro and con claims about garlic and magnets appeal in this chapter is not the laboratory-experimental kind of experience. It is a generalized claim that nature behaves in certain ways, and that a given community thinks it understands those ways well enough. What is interesting about this appeal is that it is not an explicit appeal to understanding, but hides itself, to a large extent, behind an appeal to something else, something called experience. The ambiguity that should, and does, trouble us about this is that our understanding of empiricism as

42. The terminology comes from Goodman 1955, where Goodman introduces two novel color predicates, grue and bleen, that work startlingly differently than our more familiar predicates, blue and green. The point was to problematize the question of whether particular observations of phenomena can be taken as confirming a given hypothesis: why do we think our observations confirm the statement "all emeralds are green" any better than they confirm the statement "all emeralds are grue"?

43. Goodman 1978, 101.

a method wants to suppress the theoretical commitments that so heavily inform what we think of as our experiences. The point, though, is not the old one about the theory-ladenness of observation, where theories inform how we perceive what we observe; nor is it that experience can be illusory (as in the case of hallucinations).[44] The point is instead that the theoretical aspects of this category called experience are so strong that, at their extremes, they have led the actors in this chapter to really believe that they have had experiences that are impossible. It simply cannot be the case that both sides of the garlic-magnet debate—both appealing to experience—are right. One set of these experiential claims is entirely the product of theory. Or is it both?

An important part of the point of this chapter is methodological. I have taken as my starting point a question put best by Bas van Fraassen: "Is there any *rational* way I could come to entertain, seriously, the belief that things are some way that I now classify as absurd?"[45] I have then tried to frame a way of understanding how we can deal with the many apparently—or even transparently—ridiculous claims of premodern science, and it is this: We should take them seriously at face value (within their own contexts). Indeed, they have the exact same epistemological foundations as many of our own beliefs about how the world works (within our own context).

This methodological approach sidesteps an objection that those of us who work on premodern science must still frequently deal with. The objection generally goes something like this: How is it that the practitioners of early modern/medieval/ancient science could have been so blind to the fact that theory *x*, *y*, or *z* is patently contradicted by observation (where theory *x*, *y*, or *z* is typically something in the general vicinity of astrology, alchemy, or natural magic, but can even include theories like Aristotelian uniform circular motion, which makes the objection a sweeping one)?

A good deal of work has been done at the junctures of philosophy, anthropology, and cognitive psychology to show the various ways in which classifications, theories, and kinds interact. Several things emerge: (1) classifications are "highly correlated"[46] such that features of one member project naturally onto other members of a class—a phenomenon we might think of in the present context as "veridical contagion." This has a prominent epistemological import in that it creates, among other things, the very possibility of induction, but as we have seen in this chapter, it also brings with

44. For these as standard critiques of philosophical empiricism, see van Fraassen 2002, 121f.
45. Van Fraassen 2002, 73, emphasis his.
46. Gelman and Markman 1986, 184f.

it some characteristic blind spots. (2) It is extremely difficult to pin down exactly how category formation takes place, and how members of classes get grouped together.[47] What is clear, however, is that accounts pointing to bare similarity are inadequate in all contexts.[48] Instead, classification transcends—perhaps we should even say trumps—similarity, and in fact similarity may emerge as a by-product of classification.[49] This is not to imply, however, that classification should be infinitely variable across cultures.

> The world could be partitioned in a limitless variety of ways, yet people find only a miniscule subset of possible classifications to be meaningful. Part of the answer to the categorization question likely does depend on the nature of the world, but part also surely depends on the nature of the [classifying] organism and its goals.[50]

Then again, there is enough variability for the problem to be difficult and interesting. This points to one of the most important ways in which the history of the sciences can prove particularly illuminating. By providing fully fleshed-out instances of what are really alternative classification systems, alternative kinds, and alternative criteria for kind-membership, it allows us to more fully explore the ranges and limitations of one of our most fundamental cognitive tools for approaching and understanding the world.

None of the experience claims in this chapter are disingenuous. Neither we nor Plutarch are avoiding a crucial test out of fear, credulity, or duplicity. We simply don't need to get our hands dirty. This is in part because the idea of the test becomes problematized only when we realize that there *are* conflicting claims resting on identical evidential bases—only then does a crucial test even suggest itself. Otherwise, we simply have an epistemological blind spot. At the same time, we recognize (as Plutarch did) how useful and reliable our classification systems are, and so even as the challenge is raised, we remain pretty confident, deep down, about what would happen to the magnet in our kitchen. The generalized appeal to experience has a lot of force, and it still has the power to trick us into thinking that the so-called "empirically obvious" is more properly empirical than it is just obvious.

While we have primarily focused on classification in this chapter, we have at the same time been inevitably attending to the important roles played by

47. Medin 1989; Murphy and Medin 1985.
48. E.g., Goodman 1972.
49. Medin 1989.
50. Medin 1989, 1469.

certain kinds of entities and the possibilities for interaction between them ("forces," to be a little anachronistic): magnetism, attraction, sympathy. In some contexts, though, entities, laws, and possibilities reach even further. Not only do they affect how objects in the world act, or how they stand in relation to each other and to their observers, but they may even go so far as to constrain us as agents directly, by creating, limiting, or determining possibilities for human action and for human moral responsibility, as we shall see.

CHAPTER SEVEN

The Long Reach of Ontology

In the last chapter ontology bobbed to the surface many times, but the focus of the argument there was primarily on classification. In this chapter, it is ontology's turn in the spotlight. The cast of characters and the scene have already been established: sympathy, laws, the gods, standpoints, classes. In this chapter, we will build on this to further explore the roles of some of these agents and the widespread effects that relatively simple-looking starting points turn out to have. Worlds, we might say, have a sensitive dependence on initial conditions.

We have already seen some of the long reach of sympathy as a causal force in antiquity, and we turn now to exploring one of sympathy's broadest fields of play, that of astrological causation. As we shall see over the next two chapters, sympathy was not the only way of understanding astrological causation or the knowledge of astrological effects, although it clearly forms one important touchpoint even for alternative ways of understanding the relationships between the heavens and the earth.

As historians, we have come a long way from the days when Otto Neugebauer famously attacked George Sarton in the pages of the journal *Isis* over the question of whether serious scholars should be wasting their time on "wretched subjects" like astrology.[1] By now astrology has emerged as an important minor player in the political history of the Roman empire. It has taken an even more important place in the history of the ancient sciences.[2] But gaps remain. The relationships between astrology, philosophy, and

A version of this chapter was originally published as Lehoux 2006b.

1. Neugebauer 1951.

2. On astrology and Roman imperialism, see Hardie 1986; Barton 1994a, 1994b, 1995; Potter 1994; Abry 1988, 1993, 1996; Domenicucci 1996; Ramsay and Licht 1997; Rehak 2006. On its place in ancient science, see, e.g., Barton 1994b; Jones 1999; Rochberg 2004.

physics (and in particular its epistemology) have received relatively little attention.[3] For all that, the ancient sources themselves are far from silent on the matter. Indeed, much careful thought and a considerable amount of ingenuity were deployed by ancient authors to work out the problems posed by the physics and epistemology of astrology, and its implications were widespread not just for cosmology, but even for very basic questions like whether or not human beings have free will and, bundled in with that, whether they can be said to have any moral responsibility if their actions are predermined.[4]

The reasons for this emerge readily if we perform a small thought experiment, imagining for the moment that astrology is, in point of fact, true. Let us assume for the sake of argument that the world we live in is sensitively influenced by some force exerted by the sun, the moon, the planets, and the stars—something like gravity, but with different effects. Gravity actually provides a nice parallel here, since it is something we all believe in, but something no one has ever seen directly. We are comfortably certain of gravity because we have so often seen what we take to be its effects, but of course such reasoning is not always foolproof.[5] This is not to say that we should doubt gravity—not in the least—but only that for most of us our knowledge of it has more to do with all the things we think we've seen gravity do, with the class of things we know to be affected by gravity, and with what experts have told us gravity does, than with what gravity actually is. So without yet positing a material substrate for astrological influence, let us just accept for the time being that such a stuff exists, and then look to see what it explains and how it works. On the face of it this change does not look terribly earth-shattering, but the world where such a force exists becomes interestingly different from our own.

Given this shift, the affairs of humans ranging from the outcomes of political campaigns to the successes we have in love and business are now, to a greater or lesser extent, accounted for by stellar influence. Having opened up this possibility, we simultaneously open up an epistemological possibility, the possibility of our knowing more about the future than we could have otherwise. The epistemological problems that confront us with making astrological predictions themselves, though, may be no different from those that confront the modern physicist or the biologist. If the pre-

3. Happy exceptions are Long 1982 and Barton 1994b. The most complete treatment is still Bouché-Leclercq 1899.

4. On the Stoic handling of determinism and the moral responsibility question, see Bobzien 1998; Inwood and Donini 1999.

5. A very nice treatment of both the certainty and the worry is Goodman 1972.

dictions are based to some extent in observations of past correlation, then we are faced with the same problems around induction and generalization that philosophers of science will be familiar with. To the extent that the predictions are based in an understanding of the physics of stellar influence, then the certainty of those predictions reduces to the certainty of that physics itself. In short, the strength of our belief in astrological prediction may be correlated with the strength of our belief in its physics, the sufficiency of the data set, or (most likely) some combination of both. And on the most complete view, we likely further have an understanding—and a confidence—that derives in part from the testimony of reliable authorities.

But what kinds of predictions are we talking about? On the ancient understanding, astrology covers a lot more ground than a modern newspaper horoscope does. It can account for everything from an individual's personality quirks and dispositions to large-scale political and social events, to racial characteristics, crop yields, plagues, storms, and earthquakes. Its predictive and explanatory ranges include some of what is covered by the modern disciplines of psychology, economics, sociology, medicine, meteorology, biology, epidemiology, seismology, and more. One of the major ways in which ancient astrology differs from these modern sciences, though, is in the kinds of predictions it offers, for they are not statistical. To see the force of this point, think of a modern-day insurance company. An insurance company can, depending on its data sets, calculate with a high degree of accuracy what a given individual's chances are of dying or sustaining various specific kinds of injuries in any given year.[6] Ancient astrology, on the other hand, often claims to be able to predict both how and when a given individual will actually die, even to the very hour. Not just likelihoods, but specific, datable outcomes. The belief that certain expert practitioners had access to knowledge about the deaths of important people, especially of emperors, becomes an interesting dramatic tool in the hands of Roman historians (just as divination had earlier been in the hands of Greek tragedians). On the one hand, astrology was sometimes used as a tool for the consolidation or proclamation of power. Augustus, for example, had tied much of his public image with an astrological sign of personal import, Capricorn (promoting the idea of rebirth),[7] and he even seems to have gone so far as to have published his own horoscope in full. But all the best tools are double-edged, and less flattering predictions could be circulated by an emperor's enemies.

6. On the historical specificity of this way of thinking, see Hacking 1975, 1990; Daston 1988.

7. See Barton 1995. His choice of Capricorn is interesting but not fully understood (his birth sign was Libra).

Indeed, the widespread publication of predictions of an emperor's imminent demise could, and did, lead to political instability. Legislation was repeatedly passed to try to prevent the private practice of such predictions and especially to prevent their dissemination. Suetonius preserves a wonderful story about the emperor Domitian (AD 51–96), the object of one such destabilizing astrological prediction.[8] Toward the end of his reign, when—on Suetonius' telling at least—Domitian's megalomania came to be outpaced only by his paranoia, Domitian received a report that his death had been predicted as imminent by an astrologer named Ascletario. Domitian had, in the event, been increasingly edgy over the preceding months because the sky had been abnormally active with thunderstorms, and he had also begun having terrible dreams of his own demise. He had by this point become sufficiently jumpy that he needed little excuse to execute anybody—enemy, friend, or family. Predicting the emperor's impending death, as Ascletario had, was very serious provocation indeed.

Domitian decided to kill two birds with one stone, executing Ascletario and disproving the prediction of his own pending death in the process. The method was clever. Domitian summoned the astrologer and asked him, since he was such an expert in the prediction of deaths, whether he knew what his own *exitus* would be (*exitus* meaning "death," but more literally "departure, end"). Surprisingly, given what must have been a rather overtly menacing atmosphere in the room, Ascletario said that he would soon be torn apart by dogs. Domitian took the more obvious (and quicker) expedient of having the wretched astrologer executed by soldiers on the spot—no dogs, no truth to his predictions, no imminent death for Domitian. But the astrologer got the last laugh. His funeral pyre was disrupted by a terrible storm (more ominous weather) and the body blown onto the street where it was mangled by dogs—Roman burial practices for its poorest classes meant that urban strays were not unfamiliar with such opportunities. With the inexorability of a good narrative, Domitian did not last more than a few days longer, succumbing to a murderous conspiracy. But his end did not come before Domitian had resigned himself to his fate—with an astrological prediction no less. On the evening before his death, he set aside some apples for the next day and sighed to his companions that he hoped he would survive to eat them. He then uttered his famous prediction, *ut sequenti die Luna se in Aquario cruentaret,* "that the following day the moon would be stained with blood in Aquarius," which, if one knows how to read these things, it was. Down to the very hour predicted, according to the story.

8. Suetonius, *Dom.* XV.

If we think back to the kinds of modern statistical predictions mentioned earlier in connection with insurance companies, we find an important difference between them and Domitian's predictions: when an insurance company knows someone's chances of dying, they are not calculating that individual's chances *qua* individual, so much as the chances of any individual in the given demographic. Ancient astrology, as we've seen, aspires to be much more personal, precise, and specific. It often claims that it can tell someone exactly what they are going to do, when they are going to do it, and why. It is a very powerful tool indeed. So powerful, in fact, that astrology may not leave people much room to make what they would see as their own decisions. On a strong reading of the power of the stars over human affairs, it may be the case that individuals do not have what could be considered to be free will. Accordingly, a strict determinism seems to have been associated quite commonly with astrology in antiquity.

Interestingly, though, the fatalist conclusion is not inevitable, and not every proponent of astrology was a strict determinist. There were a number of softer (and subtler) nondeterminist positions available, and it is these that I want to spend some time excavating.[9] What we shall see is that nondeterminist arguments were intimately linked with many of the standard arguments both for and against astrology. By carefully and attentively emphasizing certain aspects of the justifications for astrology, or by coopting and reworking objections to determinism, nondeterminist accounts became viable and powerful challenges to the simpler determinist accounts. As in the case of vision and perception, we shall again see a fruitful dynamic interplay between supporters of astrology and their critics.

Four Kinds of Justification for Prediction

We began this chapter by hypothetically modifying our physics on an analogy to everyday understandings of gravity in order to give astrology some tentative play. In antiquity, there were several different—if sometimes overlapping—approaches taken, of which the physical justification was only one.[10] Indeed, even as I was outlining the physical justification, I was

9. Under *nondeterminist,* I mean to include a range of opinions that agree on the basic premise that the future is not strictly determined. There are of course varying grades of nondeterminism, ranging from the belief that earthly affairs may be partly but not completely determined, to the much stronger belief that determinism in *any* degree is false (a position we might call *antideterminism*).

10. A good overview is Long 1982 . I do, however, have some strong methodological reservations about some of the details of Long's approach: "Seldom, one could say from our point of

slipping into talk of the effects of stellar influence as *visible.* This visibility points to a second possible justification, experience. As Newton claimed, we could aim to *non fingere hypotheses* and instead try to develop an account of astrology that looks to our experiences of the world in order to establish correlations without making any causal assumptions at all. Where a causal account might begin with certain posits and proceed to deduce their effects, the empirical account begins from experience and induces general correlative rules from there. Or at least it does so in principle. We shall see that experiences can also be taken as precedent-setting or paradigmatic, and so not all empirical justifications need to be enumerative or inductive.

Perhaps the most influential of the ancient causal accounts is that offered by Ptolemy in his astrological magnum opus, the *Tetrabiblos.* There Ptolemy accounts for the effects of the stars on earth in terms of the standard four Aristotelian qualities: hot, cold, wet, and dry. Mars, which is hot and dry, has a particular set of effects on the περιέχον, the *ambient* or *atmosphere* in a broad sense (somewhere between common, breathable air, and what we mean when we refer to the atmosphere of a party). For Ptolemy, the ethereal stars are known in the first instance to have an effect on the four sublunar elements in a way that will be familiar to readers of Aristotle: the motions of the surrounding ether stir up fire and air, earth and water.[11] But there is more. Almost in the same breath, Ptolemy talks of the effluence (ἀπόρροια) or radiation (διάδοσις) of the stars.[12] Beyond mere contact-stirring of one element by another, he is saying that something is quite literally flowing out from the stars, which outflow he refers to as a force (δύναμις). Finally, we get the explicit invocation of the real hero of the story, sympathy:

> ὅτι μὲν τοίνυν διαδίδοται καὶ διικνεῖταί τις δύναμις ἀπὸ τῆς αἰθερώδους καὶ ἀιδίου φύσεως ἐπὶ πᾶσαν τὴν περίγειον καὶ δι᾽ ὅλων μεταβλητήν, τῶν ὑπὸ τὴν σελήνην πρώτων στοιχείων πυρὸς καὶ ἀέρος περιεχομένων μὲν καὶ τρεπομένων ὑπὸ τῶν κατὰ τὸν αἰθέρα κινήσεων, περιεχόντων δὲ καὶ συντρεπόντων τὰ λοιπὰ πάντα, γῆν καὶ ὕδωρ καὶ τὰ ἐν αὐτοῖς φυτὰ καὶ ζῷα, πᾶσιν ἂν ἐναργέστατον καὶ δι᾽ ὀλίγων φανείη. . . . ἥ τε σελήνη πλείστην, ὡς περιγειοτάτη διαδίδωσιν ἐπὶ τὴν γῆν τὴν ἀπόρροιαν, συ-

view, have knowledge, intelligence, and rhetorical skill been more misused than in the opening three chapters of Ptolemy's *Tetrabiblos*" (178), and we are told elsewhere that Ptolemy's "ingenuity" was "misplaced" (187). Although I think Long does try to be sympathetic to his sources, the attitude betrayed in these passages does occasionally get the better of him, and, as I am arguing here, there are better methodological alternatives. A similar problem dogs Beck 2007.

11. Ptolemy, *Tetrabiblos* I.2.2.

12. *Radiation* in the sense of "spreading out in all directions."

μπαθούντων αὐτῇ καὶ συντρεπομένων τῶν πλείστων καὶ ἀψύχων καὶ ἐμψύχων, καὶ ποταμῶν μὲν συναυξόντων καὶ συμμειούντων τοῖς φωσὶν αὐτῆς τὰ ῥεύματα, θαλαττῶν δὲ συντρεπουσῶν ταῖς ἀνατολαῖς καὶ ταῖς δύσεσι τὰς ἰδίας ὁρμάς, φυτῶν δὲ καὶ ζῴων ἢ ὅλων ἢ κατά τινα μέρη συμπληρουμένων τε αὐτῇ καὶ συμμειουμένων.

> It should be very clear to all—and briefly at that—that there is some kind of power radiating and emanating from the ethereal and eternal substance to the whole of the earthly and completely changeable realm, where the primary sublunar elements of fire and air are contained and altered by the movements from the ether, and contain and alter everything else, earth and water and the plants and animals within them. . . . The moon especially, since it is closest to the earth, radiates its effluence, and most things (both inanimate and animate) are sympathetic to it and are altered by it: the flow of rivers increasing and waning with its light, the surges of the seas changing with its risings and settings, and plants and animals either completely or partly waxing and waning with it.[13]

The language here is very careful, with a recurring reference to some outflowing from the stars that accounts for earthly things co-moving (συντρέπω) with them. The preponderance of verbs compounded with the prefix *syn-* (συντρέπω [thrice], συμπαθέω, συναύξω, συμμειόομαι [twice], and συμπληρόω) is very deliberate and has the effect of portraying earthly affairs as in concert with heavenly ones. The influence or effluence is of a particular sort. It is not just a cause of any old change, but specifically a cause of *harmonic* changes (think again of the harmonics in a lyre being caused by sympathetic vibration). We will develop the idea of harmony further in the next chapter, but for now let us just note that Ptolemy in his most sophisticated experimental work, the *Harmonics,* brought together what we would call "musicology, acoustics, mathematics, physics, philosophy, and astrology."[14] Harmony was a powerful and wide-reaching set of correspondences. Ptolemy's system of astrological causation thus has an important unity beyond bald cause-and-effect statements. Rivers, plants, and animals co-wax and co-wane with the moon. And in general the hot and cold, wet and dry mixtures embodied in the planets and stars co-move the hot and cold, wet and dry mixtures embodied in all things down here on earth. It is the importance of the *co-* part of the *co-movement* that the invocation of sympathy is meant to un-

13. Ptolemy, *Tetr.* I.2.1–3.
14. Solomon 2000.

derscore, and the explanatory reach of sympathy makes it what we might call highly *symmetrical.* Nonetheless, Ptolemy's physics is complicated and nuanced. It is neither pure Aristotelianism, nor pure Stoicism, nor is it quite an intermingling of the two.[15] Instead we see the four qualities mapped onto an account simultaneously invoking sympathy on the one hand and stellar radiation on the other, which are again two different kinds of explanation not usually found together.[16]

Ptolemy simply uses what he sees as a plausible account of astral influence and treats it as a given. He does, however, recognize the potential epistemological problems when he distinguishes the inherent certainty of mathematical astronomy from the more hypothetical nature of astrology, which he reduces to a set of questions about the epistemological status of physics. He doesn't, as it turns out, have as much confidence in astrology as he does in astronomy, and this is "because of the instability and inscrutability of matter."[17] Old-fashioned historians used to pounce on these statements to try to rationalize Ptolemy by claiming that he himself downplayed his astrology in favor of his "real" sciences. Not only does this state of denial induce a deliberate blindness about major portions (and applications) of Ptolemy's project, but it also leads to a view that rather disconcertingly needs to condemn *physics* as a second-rate science at the same time as it condemns astrology.

Some ancient thinkers are even more cautious about physical and causal accounts than Ptolemy is. Sextus Empiricus goes to some length in the *Adversus mathematicos* to discount causation generally, and astrological causation specifically. What is interesting, though, is that Sextus does still allow some kinds of astrology to stand. Astrological weather prediction, which claims to be based in observation rather than causation (or to anachronize a little: induction rather than deduction), is perfectly acceptable to Sextus. He opens his attack on the astrologers thus:

> περὶ ἀστρολογίας ἢ μαθηματικῆς πρόκειται ζητῆσαι οὔτε τῆς τελείου ἐξ ἀριθμητικῆς καὶ γεωμετρίας συνεστώσης . . . οὔτε τῆς παρὰ τοῖς περὶ Εὔδοξον καὶ Ἵππαρχον καὶ τοὺς ὁμοίους προρρητικῆς δυνάμεως, ἣν δὴ καὶ ἀστρονομίαν τινὲς καλοῦσι, τήρησις γάρ ἐστιν ἐπὶ φαινομένοις ὡς γεωργία καὶ κυβερνητική, ἀφ᾽ ἧς ἔστιν αὐχμούς

15. *Pace* Tester 1987, 68f.

16. And while many modern accounts of sympathy tend to talk of it in radiative terms, there is little evidence for this reading outside of passages such as this one. To explain sympathy as it appears in most ancient sources in terms of radiation is to turn what is being offered as an *explanans* into an *explanandum.*

17. Ptolemy, *Almagest* 6.12–15.

τε καὶ ἐπομβρίας λοιμούς τε καὶ σεισμοὺς καὶ ἄλλας τοιουτώδεις τοῦ περιέχοντος μεταβολὰς προθεσπίζειν.[18]

> It now lies before us to inquire into astrology, or "mathematics"—not the whole of arithmetic and geometry taken together, . . . nor the predictive ability of the followers of Eudoxus, Hipparchus, and other such men, which some people also call "astronomy," *for this is the observation of phenomena,* as in farming and navigation, from which it is possible to foretell droughts and downpours, plagues and earthquakes, and other such atmospheric changes.

Sextus is going to take the astrologers to task, but not everyone who might be called by that name will be targeted. Some of them, the astrometeorologists, are simply correlating phenomena through observation, and that is fine by him.[19] The reason that this practice is acceptable goes back to one of Sextus' positions in the logic of signs ("signs" here meant in the ancient technical sense of evidence used in inferences).[20] In conditional statements of the form *if x, then y,* the connection between the sign *x* and its consequent *y* is acceptable for Sextus only if it is known through experience. Observation accounts are experiential (Sextus uses the technical term "reminiscent") and therefore acceptable, and causal accounts are theoretical ("indicative") and therefore uncertain and unacceptable. Sextus rejects horoscopic astrology on his contention that it is grounded in causation rather than observation.

A different worry about the physics of astrology had been earlier floated from within the astronomical tradition itself, by Geminus of Rhodes (first century BC). What is interesting is that for Geminus it is *because of physics itself* that we are prevented from offering a causal explanation of astrology. Geminus argues this by laying before us what he sees as damning evidence that should preclude astrological causation. He refers us to a story about Mount Cyllene just west of Corinth, a story that is also told about Mount Athos in the Pseudo-Aristotelian *Problems.*[21] Pilgrims sometimes venture to the top of this towering peak, 10 stades (approximately 2000 m) up, where they perform sacrifices to Hermes, including the usual burning of a sacrificial animal's thighbone. But when those same pilgrims return in later years,

18. Sextus Empiricus, *Adv. Math.* V.1–2, emphasis in translation mine.

19. On these people and their methods, see Lehoux 2007a.

20. See Allen 2001.

21. Geminus, *Elementa astronomiae* XVII.3; Aristotle, *Problemata* 944b11f. Compare also Aristotle, *Meteorologica* 340b36.

they find the ashes of their previous sacrifices perfectly undisturbed. This is because, Geminus tells us, there is no wind up that high; our pilgrims have ascended above the changeable atmosphere. But if the changeable atmosphere doesn't even reach up 10 stades, how can it possibly be affected by the terribly distant stars? The physics, he says, is impossible. It is as though the earth were surrounded by an immoveable buffer zone, so that the motions of the ethereal stars are now separated from the motions of the changeable atmosphere.

Nevertheless, Geminus tells us, we can still use signs in the sky to predict changes in the atmosphere, and past observations can be carefully collected and used to formulate predictions about the future. We may no longer have a ready answer for *why* a given star forebodes a certain wind, but we do still know *that* it does. As we saw Quintus saying in the *De divinatione: non quaero cur, quoniam quid eveniat intellego,* "I do not ask *why,* because I understand *what* happens."[22] Geminus would have been perfectly happy with Quintus' methodology.

A third justification for astrology that we find in ancient sources is, like the causal account, ontological, but rather than positing physical forces to account for stellar influence, it posits theological ones. It invokes the gods as real causal agents in the world. They become actors in the drama of the physical world, and their interest in our welfare has them sending us important messages in the stars. The Stoics put it best in an argument we saw in an earlier chapter:

> If there are gods and
> (a) they love us, and
> (b) they are not ignorant of the future, and
> (c) they know that knowledge of the future would be useful for us, and
> (d) it is not beneath their majesty to communicate with us, and
> (e) they know how to communicate with us,
> then divination [of which astrology is one type] exists.[23]

This is rather a more exhaustive spelling out of an implication that is usually shortened to "If there are gods then there is divination," which version depends on a conception of deity that implies all of (a) through (e), a concep-

22. Cicero, *De div.* I.15.

23. This formalizes the argument in Cicero's *De div.* I.82–84. Cicero also tells us that Panaetius was one of the very few Stoics who rejected astrology, but his reasons for doing so are unknown (*De div.* II.42).

tion that is more or less common in antiquity.[24] As far as the big premise goes, the existence of gods, this was taken as given by nearly everyone in antiquity, as we've seen.

Not only do the gods guarantee the truth of astrology, but, in the first-century Stoic author Marcus Manilius, they also initially seem to reveal the secrets of the heavens to the first investigators:

> [sacerdotes] . . . quibus ipsa potentis
> numinis accendit castam praesentia mentem,
> inque deum deus ipse tulit patuitque ministris.
> hi tantum movere decus primique per artem
> sideribus videre vagis pendentia fata.[25]

> The very presence of powerful divinity kindled a pure mind for the priests, and god himself brought them to god and revealed himself to his servants. These began our honored study, and by that art were the first to see the fates that dangle from the wandering stars.

Manilius tells us that the diligent and god-inspired investigators then went on to use careful observations over long periods to develop the science of astrology fully, so astrology gets its veracity through an interesting and unique combination of empiricism and divine revelation.

As for Sextus, Geminus, and Cicero's Quintus, the observations that enable astrological prediction to function in Manilius are long-term, multiple observations that build up knowledge over time. We tend to forget, though, that not all empirical justifications work this way. In some cases, single experiences can be sufficient to convince us that a phenomenon or a law is true. We saw in the case of Domitian, above, that the confluence of coincidences was sufficiently implausible, when Ascletario's body was set upon by dogs in the street, that Domitian was turned from a sceptic to a believer in an instant. To draw a simple modern example, perhaps we would be similarly convinced that a lottery was rigged if an accuser correctly named the winner in advance—much hinges on inherent plausibilities, though, and

24. See Cicero, *De div.* I.9–10. The argument was widely, but not universally accepted. Holdouts included (1) the Epicureans, who would deny pretty much all five of the qualifications, and (2) those for whom clause (c), was seen as a potential weak spot for the deterministic Stoics, since a revelation of the inevitable future would not allow us to avoid that future, in which case what good is it? (It is just this dilemma that the Oedipus story plays on.)

25. Manilius, *Astronomica* I.48–52. Manilius' Latin is exceedingly complex and famously difficult to translate.

it is here that the modern sceptic wants to dismiss Domitian as jumping to the wrong conclusion. Peter Galison has drawn attention to the use of single images in modern microphysics as proof of the existence of new phenomena or of new entities, as when a new particle is finally demonstrated through a single but distinctive event in a cloud chamber. Even today, one good picture of a single paradigmatic event can do more epistemological work than thousands of statistical data points under certain circumstances.[26] Finally, although I am not aware of examples from classical astrology directly, it was not uncommon in antiquity for particularly important events to color calendrical dates in perpetuity, as was the case with the Ides of March, which was always seen as a bad-luck day after the murder of Caesar on that date in 44 BC.[27] This kind of case is especially interesting as it seems to take the evil event as precedent-setting. Outside of the classical world, we can point to Mesopotamian astral divination, where ideas of astrological precedent were highly judicial,[28] but, granting the important roles Mesopotamian astronomy and astrology played in the development of the Greco-Roman astral sciences, whatever judicial features we might see in their epistemology were culture-specific and do not seem to have been carried over with the other aspects of astrological doctrine and technique that came to the Greeks and Romans from Babylon.

Thus far we have four different models for grounding astrology in antiquity: physical, inductive, theological, and paradigmatic. Because each of these is an epistemology for predictability generally, each has important, and importantly different, implications for our understanding of our relationships with the predicted future. More to the point, each of these epistemologies has different implications for, and different ways of dealing with, questions of human free will and determinism. To be sure, the sharpness of the distinction between these models is a function of the critical distance at which we stand.[29] In the first instance, the fourfold categorization I am offering is nowhere made explicit in my sources, and sometimes the distinctions are less sharp than I am portraying them here (see my comments on Seneca's and Ptolemy's observational accounts, below). The point of the models I argue for here is to draw out some fruitful distinctions for exam-

26. Galison 1997. Galison's qualification that such images in microphysics are *mimetic* need not worry us here. My point for the moment is just about the epistemological category of paradigmatic instances.

27. See Grafton and Swerdlow 1988.

28. I flesh this point out more fully in Lehoux 2003b and 2006b.

29. I would like to thank Thomas Laqueur for drawing my attention to this set of points and for encouraging me to elaborate on them.

ining the interrelationships between epistemologies, ontologies, and larger theoretical developments in general. It may be the case that if we stand at an even further remove we may see that some grounding assumptions are shared by several of my epistemological camps. One is the agreement that the universe is nonrandom. Even if the gods may look to be acting from our point of view in some inscrutable way, they themselves—their existence—remains a constant. It is not as though they may wink out or be replaced by different kinds of entities at any moment, and the assumption always is that they have their reasons for doing what they are doing. Further, we can see close relationships between the Ptolemaic and the Stoic accounts, both of which invoke causes we would class as physical, and they share an emphasis on sympathy and antipathy in particular. The theistic and empirical accounts may also be seen as sharing a bipolar heavens/earth superstructure. Without retreating to a simplistic "as above so below" position, we can see that there are semiotic similarities in the theistic and observational justifications. A bare theistic account, for example, does not necessarily have direct Ptolemy-type physical causation whereby the stars instigate a causal chain of events. (Although the Stoic model often will run in this direction, it is importantly loaded with extra assumptions about the unity and beneficence of the cosmos as a whole.) Like the bare observation account, bare theism may only imply the correlation of events as signs rather than tying stellar and earthly events into a simple and direct causal chain. The difference, though, is that bare observation accounts aim to make no causal assumptions at all, whereas the theistic accounts will run to some version of third-party causation. So, for example, on the simplest version of the Stoic theistic account, the gods send signs because they know of future events about which they wish to warn us. At the more complex end, the world itself is an ordered divinity, and the signs emerge as part of the same unfolding of an inevitable causal chain that also brings about the foretold events. The causation then becomes part and parcel of the constitution of the universe, a constitution that is itself the very definition of a divine ordered rationality.

Predictability and Determinism

Of all the predictive sciences in antiquity, astrology is especially prone to questions of determinism. This is because astrological signs are importantly different from other kinds of *omina*. Omens from the flights of birds, or from marks in a sacrificial liver, happen in such a way that the diviner never really knows what signs to expect or even in many cases when they might appear. Astrologers on the other hand are faced with the unique situation that all of the signs from which they make their predictions are themselves predictable and regular. The astrologer knows in advance what the positions of the stars are going to be for any given day, and can deduce from that data what the earthly outcomes are going to be. On the face of it, the future may just all be fallout from the original constitution of the cosmos, and the astrologer can effectively look at any point in time from any point in time as though it were written out in front of him.

Accordingly, many classical sources straightforwardly equated belief in astrology with belief in determinism. As Tacitus said: "Most men do not doubt that what will happen to them is predestined from birth," and the blame for any incorrect predictions can be laid squarely on the shoulders of the inept practitioner.[30] One of the most interesting and colorful arguments for this determinism comes in Firmicus Maternus' triumphalist description of the death of the Neoplatonic philosopher Plotinus, where Plotinus is forced to revise his rejection of fate when the goddess Fortune, personified and brutally vindictive, teaches him a hard lesson:

> ecce in quadam parte orationis suae, sicut mihi videtur improvidus et incautus, vim fatalis necessitatis adgreditur et homines Fortunae decreta metuentes severa orationis obiurgatione castigat, nihil potestati stellarum tribuens, nihil fatorum necessitatibus reservans, sed totum dicens in nostra esse positum potestate. . . .
>
> ecce se illi in ista confidentiae animositate securo tota fatorum potestas imposuit et primum membra eius frigido sanguinis torpore riguerunt et oculorum acies splendorem paulatim extenuati luminis perdidit, postea per totam eius cutem malignis humoribus nutrita pestis erupit, ut putre corpus deficientibus membris corrupti sanguinis morte tabesceret; per omnes dies ac per omnes horas serpente morbo minutae partes viscerum defluebant, et quicquid paulo ante integrum videras, statim confecti corporis exulceratio deformabat. sic corrupta ac dissipata facie, tota ab illo figura corporis recedebat et in mortuo, ut ita dicam, corpore solus superstes retinebatur animus, ut ista gravis morbi continuatione confectus et tormentis propriis coactus ac verae rationis auctoritate convictus vim fati potestatemque sentiret et ut confecti corporis laceratione quassatus sententiam Fortunae pronuntiantis exciperet.[31]

30. Tacitus, *Ann.* 6.22.
31. Firmicus Maternus, *Mathesis,* I.7.18–21.

> Look how in one part of his work [Plotinus] attacks the power of the necessity of fate (quite foolishly and carelessly it seems to me) and he forcefully rebukes people who fear the decrees of Fortune. He grants no power to the stars, and he offers no necessity to fate, but says that everything is within our power. . . .
>
> And look how, when he was secure in this impudent rashness, the power of fate compelled everything: first his limbs became stiff from a chilling and torpor in his blood, and the sharpness of his eyes slowly lost their clarity as the light in them failed. After this, his whole skin erupted in a pestilence fed by malignant humors, so that his putrid body melted away into death with soured blood, failing limbs. Every day and every hour small parts of his viscera were dissolved by the creeping disease, and what was seen as intact one moment was deformed the next by the ulceration destroying his body. Thus corrupted and dissolved in appearance, the whole shape of his body fallen apart, all that remained in the—so to speak—dead body was the mind, so that being destroyed by the horrible progressing disease, he was convinced by his own torments and by the authority of true reason to see the force and power of fate. Thus broken and with a mangled, destroyed body, he received the sentence passed by Fortune.

Perhaps partly because of the absence of statistical accounts, predictability and determinism are often seen to be coentailed in antiquity. This is certainly true of the theistic Stoic position as sketched by Manilius, for example. While other astrologers were more than happy with a strict determinism, not everyone else would be.[32] Accordingly, various ways *out* of the determinism potentially inherent in a true astrology were on offer, with these different ways out dialectically engaging in part with the different epistemological justifications for astrological prediction. At the same time, though, attempts to reconcile some degree of free will with some degree of predictability also embrace and rework objections leveled by astrology's critics.

Physical Solutions to Determinism

For the physical reading of astrological prediction, where the stars themselves have a direct effect on earthly affairs, attempts to reconcile nondeter-

32. Indeed, many Christian rejections of astrology were just rejections of determinism. The way around this objection was, as Manuel Comnenus saw (*CCAG* I.106f.; compare Tester 1987, 95) and Augustine alludes to, seeing the stars as mere signs, rather than causes.

minism with predictability tend to cluster around the physics or the epistemology of that physics. In the case of Ptolemy, we see him give some careful consideration to the conditions under which reliable prediction would be possible. He treats these conditions hypothetically: What, he asks, would prevent someone who accurately knew

(a) the exact and minute motions of the stars,
(b) the nature of each of their particular effects, and
(c) the effects of their combinations,
from (d) predicting the particular conditions of the atmosphere?

But notice here that it is not yet the fates of *humans* he can predict reliably. It is the conditions of the atmosphere, the ambient. To get to human personalities, complexions, and dispositions, we need to go one step further and formulate arguments about how the physics of the atmosphere interacts with the physical composition of people. Such arguments are not far to find in ancient sources generally (the idea was a cornerstone of ancient medicine), nor in Ptolemy in particular, but the double level of causation sets the final human predictions at two removes from the original signs.

The possibility of acquiring all the requisite knowledge is an important qualifying factor. Ptolemy tells us immediately how difficult this is in practice. The science, he says, is very large, complex, and heterogeneous, and because it deals with physical matter, can only attain to a limited degree of certainty. Thus there are some things that by their very nature simply cannot be predicted due to the inherent uncertainty of matter: "The whole science of the qualities of matter is conjectural rather than certain" (ἡ περὶ τὸ ποιὸν τῆς ὕλης θεωρία πᾶσα εἰκαστικὴ εἶναι καὶ οὐ διαβεβαιωτική).[33] Matter is "flimsy and uncertain" (ἀσθενὲς καὶ δυσείκαστον),[34] and this is no minor point for Ptolemy. It is just the dependence of astrology on the uncertainties of physics that steps astrology down a rung from astronomy as a science. Where the truths of astronomy, being mathematical, are certain and knowable, the truths of astrology, being physical, are merely probable.[35] Note,

33. This is a slight modification of Ptolemy, *Tetr.* I.2.7; the grammatical cases have been changed to maintain sense.

34. Ptolemy, *Tetr.* I.1.1.

35. Older versions of the history of astronomy tended to make great hay of Ptolemy's separating his astronomy and astrology into two books (the *Almagest* and the *Tetrabiblos*), as though that pointed to doubts Ptolemy had about astrology as a body of knowledge. But Ptolemy is clear that even if it is less *certain*, astrology is more *useful* than astronomy. See also Bowen 2007.

though, that Ptolemy's epistemological point about physics is not the same thing as a modern scepticism about astrology. Modern sceptics say that astrologers are too much like fairground psychics. Ptolemy is saying they are too much like physicists.

The other salient factor in Ptolemy's nondeterminist account has to do with the second, mundane half of the two-tier astrological causation model. Although the atmosphere is known to have an effect on people, it is not, Ptolemy reminds us, the only factor. There are many other causes at work in the developing of a given person's constitution. These συναίτια, these "co-causes," combine with the astrological causes to produce specific effects. Ptolemy cites the qualities of the seed of the parents and the geographical location as two of the primary co-causes that determine the character of an individual.[36] Pliny likewise invokes the complexity of causal networks as a qualifying factor when emphasizing the important effects of the stars on the weather.[37] The stars, then, are causes, but only one set of causes among many.

Ptolemy's explicit mention of the importance of geographical location bears some elaboration. Geographical variation is commonly invoked in astrological texts as a basic *explanandum.* The idea here is that astrologers need to explain why people born in one country are so similar to each other and so different from everyone else. Why are the Ethiopians all so dark and the Scythians all so pale, if it is the same conjunctions of the same stars that are responsible for human complexions and characteristics? In a marvelously self-serving passage Firmicus Maternus has an imaginary interlocutor frame the question thus:

> si Saturnus facit cautos, graves, tardos, avaros ac tacitos, Iupitter maturos, bonos, benignos ac modestos . . . cur quaedam gentes ita sunt formatae ut propria sint morum quodammodo unitate perspicuae? Scythiae soli immanis feritatis crudelitate grassantur, Itali fiunt regali semper nobilitate praefulgidi, Galli stolidi, leves Graeci, Afri subdoli, avari Syri, acuti Siculi, luxuriosis semper Asiani voluptatibus occupati, et Hispani elata iactantiae animositate praeposteri.

> If Saturn makes people prudent, serious, slow, greedy, and quiet, and Jupiter mature, good, kind, and moderate . . . why are some groups constituted so as to have particular common characteristics? The Scythians

36. Ptolemy, *Tetr.* I.2.8.
37. Pliny, *Nat. hist.* II.105.

> alone attack with beastly and savage cruelty, the Italians show an ever-noble regality and glory, the Gauls are thick, the Greeks effeminate, Africans cunning, Syrians greedy, Sicilians sharp-witted, Asians always preoccupied with luxurious pleasures, and Spaniards foolish with ridiculous boastfulness.[38]

For Ptolemy the explanation is simple: geography and local climate are important co-causes of personal characteristics, ones that astrological causation works with to make one Scythian more or less cruel than another Scythian, one Roman more or less noble than another.

What is interesting here are the different roles geographical variance plays. In Ptolemy it arises as a *qualification* to strict determinism, whereas in Firmicus Maternus it comes in as a difficult *objection* to Firmicus' strictly deterministic astrology (its difficulty is shown by the number of feints Firmicus makes at answering the objection before finally looking it in the eye at the very end of book 1). This coopting of objections to deterministic astrology by nondeterminist astrologers like Ptolemy is not uncommon, and it is rhetorically quite powerful.[39] (One similarly wonders whether Ptolemy's use of the double layer of causation, where the stars in the first instance affect the atmosphere, τὸ περιέχον, may not likewise have something to do with Sextus' acceptance of τοῦ περιέχοντος μεταβολὰς προθεσπίζειν, "the prediction of changes in the atmosphere.")[40]

We see this retooling of objections to astrology being put to powerful use by Seneca in the *Naturales quaestiones*, in a passage slightly but interestingly different from the multiple causation arguments of Pliny and Ptolemy. The difference in Seneca is that instead of appealing to the multiplicity of earthly causal forces, he cites the multiplicity of astrological causes as leading to uncertainty about the future and inaccuracy of prediction.[41] Where opponents of astrology were fond of parading famous mistaken predictions, Seneca preempts that move by admitting that mistakes not only *can* be made, but *must* sometimes be made. However, these are mistakes of interpretation only, and this raises an important point: we may not have complete predictive command of all the myriad effects of the stars and their combinations, but the effects are there nonetheless. Where in Ptolemy and Pliny the effects were moderated by external (i.e., nonastrological) causes,

38. Firmicus Maternus, *Math.* I.2.2–3.
39. See also Long 1982.
40. Compare Ptolemy, *Tetr.* I.3 et passim with Sextus, *Adv. math.* V.2.
41. Seneca, *Nat. quaes.* II.32.6–8.

Seneca is saying that the internal effects are all-important, but impossible to control exhaustively. Seneca is a determinist, but at the same time he is a determinist who is less confident about our knowledge of the future than Firmicus Maternus was. This brings up an important point: the physical astrologies of Ptolemy and Pliny can be nondeterministic only if the extra sets of causes are entirely independent of the stars. For Stoics this simply cannot be the case physically, due at one end of the stick to the fundamental interconnectedness of all things, and at the other end to their firm belief in fate.

The second set of justifications for astrology—observation and experience—picks up on this question of the knowability of the future. If we know the stars through past observation, then how much certainty do we really have about the future? Asking this question in an ancient context is not to raise David Hume's problem of induction, however. Instead, we see general questions about the acceptability of observation framed in terms of kinds of sign inference, or (as we have seen) in terms of the rhetorical handling of witnesses. We have already seen that observation accounts do often play a role in justifications of astrology, sometimes on their own (as in Sextus and Geminus), and sometimes in conjunction with other arguments (as in Manilius' combination of divine revelation and human investigation). Ptolemy offers us yet another variation:

> ἔτι καὶ τοῖς παλαιοῖς τῶν πλανωμένων συσχηματισμοῖς, ἀφ' ὧν ἐφαρμόζομεν τοῖς ὡσαύτως ἔχουσι τῶν νῦν τὰς ὑπὸ τῶν προγενεστέρων ἐπ' ἐκείνων παρατετηρημένας προτελέσεις, παρόμοιοι μὲν δύνανται γίνεσθαι μᾶλλον ἢ ἧττον καὶ οὗτοι διὰ μακρῶν περιόδων, ἀπαράλλακτοι δὲ οὐδαμῶς, τῆς πάντων ἐν τῷ οὐρανῷ μετὰ τῆς γῆς κατὰ τὸ ἀκριβὲς συναποκαταστάσεως, εἰ μή τις κενοδοξοίη περὶ τὴν τῶν ἀκαταλήπτων κατάληψιν καὶ γνῶσιν, ἢ μηδ' ὅλως ἢ μὴ κατά γε τὸν αἰσθητὸν ἀνθρώπῳ χρόνον ἀπαρτιζομένης, ὡς διὰ τοῦτο τὰς προρρήσεις ἀνομοίων ὄντων τῶν ὑποκειμένων παραδειγμάτων ἐνίοτε διαμαρτάνεσθαι.[42]

> And furthermore, the observations of the ancient configurations of the planets with which we accommodate similar ones obtaining now to the outcomes seen by them under the earlier circumstances, can at long intervals be more or less similar but never identical, because the precise arrangement of everything in the heavens relative to the earth either never happens, or never happens on a human time scale, unless one pre-

42. Ptolemy, *Tetr.* I.2.7.

> tends to know and comprehend the incomprehensible. Thus predictions, differing from the established paradigms, are sometimes given the lie.

In part, this claim works because of the very high degrees of accuracy to which Ptolemy wants to hold astronomy and astrology. Sloppier astrologers, the kinds casting inexpensive horoscopes in markets and temples all over the empire,[43] may be happy with a simple "Mars is retrograde in Taurus," but Ptolemy is here emphasizing that the latitude, longitude, speed, distance, and direction of Mars in combination with those of all the other planets will, if we are to speak precisely, never recur exactly identically. Worse astronomers may say so, but Ptolemy has a rather large and detailed technical command of astronomy (the *Almagest*) to back him up. The devil, he knows, is in the details.

But this is not really a nondeterminist argument. It is more like Seneca's argument for an imprecise astrology than it is like Ptolemy's other, causal, arguments for a nondeterminist astrology. In later discourses we do sometimes see nondeterminism attached to an idea of the stars as signs rather than as causes, but this is not what Ptolemy is up to here.

The Cascading Effect

Clearly, ancient astrology is far from monolithic. We have in this chapter explored one aspect of the epistemology of astrology, the interrelationships between the various positions on determinism, and the various justifications for the efficacy of astrology. By suspending our disbelief for a little while, we have allowed the ancient arguments for astrology a little bit of healthy breathing room. The upshot is this: if we admit new entities, particularly entities like gods and new physical forces, into the world, we simultaneously admit new and very far-reaching possibilities, and even new necessities. There is a rippling effect here, and one that does not stay confined to ontology. Different entities can cause the world to behave in radically different ways: *ontologies cascade*, and this in complex ways. An important way in which many treatments of astrology have failed us is in their tendency to overemphasize a disjunction in the methodology and epistemological standards across the great modern-premodern, or science-pseudoscience divides. Imagining that astrology is true is not the same thing as pretending to be foolish, credulous, or superstitious, or as completely ceasing to be

43. On the wide spread and various venues of astrology, see Evans 2004, Barton 1994b. For examples of actual horoscopes, see Neugebauer and van Hoesen 1959; Jones 1999.

rational for a moment. Astrology is, in the ancient discourses, both highly rational and eminently empirical. It is surprising how much evidence there was for it, and how well it sustained itself in the face of objections (and for a very, very long time at that).

I have also tried to work around, for the moment, readings of astrology that account for its ancient popularity using psychological, semantic, or social explanations, explanations along the lines that astrology filled a need for comprehension and control of the chaotic universe in premodern societies; or else it only *seemed* true because the predictions it made were vague enough to be polyvalently adaptable to any outcome; or else it facilitated professional status distinctions and social control mechanisms for learned elites over the ignorant masses.[44] Some of these points may have some truth to them (although I argue in the next chapter against polyvalent vagueness), but at the same time they ignore much historical evidence as well. Defenders of astrology often wielded formidable arguments that need to be taken very seriously if we are to fully understand the roles of astrology in the worlds in which it operates. The fact is that most ancient thinkers who talk about it seem to think that astrology really did work, and this for very good reasons. Future physicists may well look back at the baby steps of the simple matter theory of the twenty-first century. But their accounting for my belief in twenty-first-century physics solely in terms of sociological control mechanisms or the psychological need for me to impose desperate order on a chaotic universe would be to miss what I see as the major reason for accepting our physics as true. I believe it accounts for virtually all the phenomena I have seen (or at least paid a certain kind of attention to), and I am told that our best minds, given the best available tools, agree. Obviously, it cannot account for all the phenomena all the time, but—as for the ancient astrologers—it does so more than enough to be thoroughly convincing as an explanation.

44. I would like to thank Peter Struck for this threefold categorization. For a recent example of the "filling a need for comprehension and control" thesis, see Maul 1999.

CHAPTER EIGHT

Dreams of a Final Theory

Explaining the Cosmos

Consider the common suffix *-ist*. When we use it in politics, religion, art, or philosophy, an ist is generally correlatable with an *ism*. So Marxists advocate Marxism, realists realism, anarcho-primitivists anarcho-primitivism. In the sciences, on the other hand, we don't tend to have such isms for our ists. There is no biologism, no chemism, no solid-state physicism, no economism, at least not in the school-of-thought relation that we find in the isms of politics, religion, art, or philosophy. Why is that? As a first approach, we might suspect that isms have more to do with belief and choice, and the scientific ists have to do with professions and fields of research. From an etymological point of view, I'm sure there are good historical reasons for our scientific ists, but for the present purposes it is enough just to draw attention to the fact of an apparent (or at least easy) point of difference.[1] This modern ism-lessness marks a point of contrast with how we understand the ancient sciences, where our various sources are so often parsed and understood just exactly in terms of their isms: their Stoicism, their Platonism, their Epicureanism. If we see isms as being about belief choices, or about deliberate stances in arguments with other traditions, then we may have a correlative tendency to think of the relationship between observation and theory in antiquity as being differently mediated, filtered, or distorted by beliefs about the world than modern scientific theory is. Ancient science may just look like the development and defense of systems for systems' sake.

There is a second problem with overattention to isms. As we've seen again and again, the boundaries between ancient schools are often permeable. Seneca is a self-professed Stoic, but then he is critical of many of his

1. See Proctor 2007, for an interesting discussion of the matter. From the epigram to Proctor's paper, I could here quote Eduard Bernstein: "Kein Ismus ist eine Wissenschaft."

Stoic predecessors, and feels at all times free to offer new solutions to their problems. Cicero is a self-professed Platonist, but feels free to borrow heavily from Stoic theology and ethics. Galen wrote a book arguing that Hippocrates and Plato together represented a complete pinnacle of truth, but then he went and extensively incorporated Stoic mechanisms and terminology into his visual theory and psychology. Ptolemy has proven impossible to pin down under one school, using elements of Stoicism, Aristotelianism, Hipparchism, and more. This kind of free and easy flexibility of allegiance, so very characteristic of the period and culture we have been studying, has also been handled by modern historians under the auspices of yet another ism: *eclecticism.*

In 1988, Dillon and Long dedicated a very useful volume to the question of eclecticism, partly to rehabilitate the idea of eclecticism (which in philosophical circles can be seen as pejorative), and partly to draw attention to what they saw as a widespread philosophical mentality, very prominent from about the first century BC through the second century AD—in other words, just the period most closely covered in the present book. Though such eclecticism was clearly prominent, there are two schools that stand as outliers to the overall trend, one (Pyrrhonist Scepticism) because the core of its philosophy was constituted in a radical rejection of even the basic starting points of most other schools, and the other (Epicureanism) because its idiosyncratic physics, hedonist ethics, and widely objectionable theology were so radically incongruous with those of most everyone else.

Looking to the traditions that turn out to be more intercompatible, those of Platonism, Stoicism, Aristotelianism, Pythagoreanism, plus many of the medical traditions that Galen and others borrow so heavily from, we find that there are two ways of looking at what is called their eclecticism. We could see eclecticism as being constituted in individuals stepping out of the various schools to borrow little bits they find useful from each other (in a kind of importation-of-nonlocal-resources model). Or—and this was the general consensus in the Dillon and Long volume—we could make a case that there is instead a wide body of agreement on fundamental issues, physical, theological, logical, and ethical, and that an individual's particular philosophical affiliation comes in as a kind of aftereffect of working out the details. In some ways, philosophical affiliations may even have more to do with what would be better called styles of doing philosophy than with specific doctrinal commitments in this period.[2] The picture of the Roman

2. See Sedley 1989; Long 2003; Gill 2003. On "styles" in a broader sense, see Hacking 2002, chaps. 11 and 12; Crombie 1994. Contrast this reading with, e.g., Schiavone's characterization of

sciences that I have been arguing for in this book has been one that agrees in broad outline with this view, where there is a good deal of agreement among many Romans about how the parts of the universe interrelate at the broadest level, about what it means to explain or to know anything about nature, and about how human ethical duty is linked to that knowledge. By focusing not on school orthodoxy and heterodoxy but instead on shared approaches to problem sets we can see a good deal of common ground between what we might instead call the "concentric" schools, which is to say, among the majority of elite, educated, and philosophically literate Romans.

In this chapter I argue that, from the standpoint of the methodology of the sciences, one of the foundational beliefs shared by many schools is the conviction that the universe is what I will call *symmetrical,* which is to say that ways of explaining parts of it will also explain other parts, and that it is symmetrical across physical, ethical, theological, and psychological dimensions. In Epicureanism the idea manifests in the use of mechanical interactions between atoms as a way of explaining a wide variety of apparently unrelated phenomena, from rocks to erotic impulses. Similarly, we've already seen pneuma get picked up by many thinkers (many of them Dillon and Long's eclectics) to explain a wide range of phenomena. In many contexts, though, symmetrical explanations can become considerably more abstract, more mathematical. This sounds familiar enough: when we apply the Fibonacci series to both celery and snail shells, we are applying one measure to two distinct phenomena—*sym-metry* in the etymological sense of co-measuring. So perhaps there should be nothing at all surprising in an ancient belief in symmetry, then, except that they sometimes apply it in very odd-looking (and very far-reaching) ways. This has had the effect of making many of the broadest ancient explanatory categories look all too broad to modern commentators. Accordingly, modern historians—even sympathetic ones—often categorize the most general levels of ancient explanations as "mystical" or "numerological." I am thinking here of, say, descriptions of planetary motion in terms of musical harmonies, or of both musical harmony and planets together in terms of number theory. When the human soul begins to be implicated in these arguments, we see (yet again) the now familiar overlapping of ethics, politics, law, and nature in Roman science. Beginning, then, with a look at the relationships between politics and the cosmos as a whole, as instanced in Cicero's *Dream of Scipio,* I argue in

Aelius Aristides: "We can glimpse the elements of a redundant and troubled eclecticism, crystallized into a somewhat feverish Platonism with occasional shadings of Pythagorean contamination (a frequent combination)." Schiavone 2000, 10.

this chapter that symmetrical explanations, even apparently numerological ones, should be rethought as vital parts of the Roman projects of knowing the cosmos, and I try to show how modern arguments for the importance of *breadth* in scientific explanations can make space for understanding the place and force of these ancient orderings of nature.

We can begin by thinking about what it would mean to ask whether there were one overall conceptual scheme, one theory, one final set of laws, to which the most important natural processes could be reduced or accommodated. What would such a theory look like? What phenomena, what whole fields, would it need to unify? Physics? Physiology? Philosophy? We've seen how broad a conception nature is for the Romans, and how firm certain lines are between the natural, the social, and the ethical. So how extensive should our broadest explanations be? Related to this: should we be worried that too much breadth of extension also means explanatory vagueness, deliberate blindness, or uncritical handling of data?

In looking at the explanatory and synthesizing roles played by the layering of structures from one set of phenomena onto another, we find that such layering produces both coherence and connectivity among phenomena. Order that repeats in multiple directions, across very different kinds of explananda paints a picture of a universe that is not just vertically reflective, but multidirectionally symmetrical. It is this symmetry that not only allows one aspect of the universe to reflect another aspect—whether it is humans and the zodiac or the elements and the seasons—but that allows the known truths of one branch of knowledge to reveal hidden truths of another. This is the inductive stretch of classification relationships, one of the most powerful of the concomitant effects of similarity, and also the proper home of similarity's explanatory power. At one and the same time, though, these known truths from one branch not only reveal, but also constrain the knowledge of other branches. This is the force of the common criticism that preexisting theories can blind us to anomalous observations—the provision of handy explanatory structures is sometimes just too handy. Symmetrical arguments thus have real benefits, and they also have known epistemological dangers. Where does the use of cosmic harmony stand?

From the earliest years of Greek philosophy, Pythagoreans and their successors had been engaged in a project of organizing and understanding the cosmos numerically. By the time the Romans encountered this material it had been honed and developed, refined and recontextualized. The sciences of harmonics, astronomy, astrology, and of mathematics had changed considerably since the days of the Pythagoreans or Plato. Number not only persisted as a conceptual tool, but found broader and broader application, deeper and deeper interconnection and empirical confirmation. By the time of Cicero, the numerical analysis and reduction of phenomena had become for many a primary tool for understanding both nature's regularities and the recurrent structures underlying the organization of the cosmos as a whole—in short, for explaining the cosmos.

I am using the term *explanation* here in something like Philip Kitcher's sense, where an explanation is one of a related family of sentences that "collectively provide the best systematization of our beliefs."[3] As Kitcher expands upon what this means, he says that "science advances our understanding of nature by showing us how to derive descriptions of many phenomena, using the same patterns of derivation again and again, and, in demonstrating this, it teaches us how to reduce the number of types of facts we have to accept as ultimate (or brute)."[4] One important part of Kitcher's definition, though, is this word "derivation," and it is there if anywhere that the acceptability of my use of explanation will hang. Part of the problem with making the argument in many Roman contexts is that the Romans were of course not looking for a comprehensive theory of explanation itself—that question is very much a product of twentieth-century philosophical concerns. Not all ancient sources took pains to explicitly lay out every step in the explanatory chain from propositions about order in the cosmos generally down to specific observable phenomena. This is not be surprising. To take a modern parallel, an answer like "this happens because of natural selection" would count in many contexts as an explanation without further elaboration. Only in certain contexts do we get explorations of what "natural selection" really means as an explanation. Indeed, much of its force is due to the bare fact of our agreeing that it *is* an explanation, never mind whether we fully understand what "natural selection" means in detail (a situation analogous to Ptolemy and Galen's for like-affects-like).

As we look to Cicero's *Dream of Scipio,* we find several prominent kinds of explanation being used to bring together politics (or ethics more broadly) and the natural world. The most difficult of these, from the point of view of trying to understand it as an explanation, is Cicero's use of number. "Seven" does a good deal of constructive work in the *Dream.* This has often been parsed by historians as a species of "numerology," nothing more than quasi-religious or mystical hand-waving. I argue instead that the ways in which

3. Kitcher 1989, 430. See also Friedman 1974. For one of the foundational views of explanation, see Hempel and Oppenheim 1948. For a handy summary of critiques and alternatives, see Kitcher and Salmon 1989.

4. Kitcher 1989, 432.

numbers (particularly harmonic numbers) are being used, the ways in which they are seen to reveal truths across a disparate range of kinds of phenomena, and the ways in which harmonic numbers frequently stand in this and other sources as satisfying ends to chains of why-questions—all of these indicate that harmony is doing important kinds of explanatory work. In some sources, most notably Plato's *Timaeus,* Macrobius, and book 3 of Ptolemy's *Harmonics,* we get fairly clear attempts to formulate and formalize a deductive chain from "the cosmos is divine and rationally ordered" through propositions about numbers and their harmonic interrelations down to why certain numbers manifest in certain phenomena (why there are seven planets, e.g.). In other sources, such as Cicero, we see the deployment of parts of such arguments, but not a complete elaboration of all the points on the way. This is not because the doctrines of number are inherently vague, but instead, as I hope to show, because of Cicero's confidence that "harmony" would be taken—without further elaboration—as a sufficient explanation by his audience.

Orbs, Souls, Laws

In the introduction to this volume I referred to the intertwining of nature, the gods, and politics as a threefold cord in Roman thought. Here we complete our historical investigation of this great braid by looking at how *explanations* interact between the three. To see how this works, there can be no better place to start than with Cicero's *Dream of Scipio,* one of Western literature's most enduring and influential short prose works. Of course, the *Dream* was not originally written as a short prose work, but is instead a fragment of a much longer dialogue, *The Republic.* The *Dream* is a fragment that was cloven off some time in the early Middle Ages not only from the rest of the *Republic* but from much else in ancient science as well. This independent survival gave it an importance and scope inversely proportional to its meager length. The vicissitudes of textual survival from Rome into the early part of the European Middle Ages highlight profound religious, intellectual, institutional, and educational changes.[5] The case of the *Dream* is particularly interesting here, because its survival was quasi-parasitic: it was in large part because of philosophical and theological interest in Macrobius' Neoplatonic *Commentary on the Dream of Scipio* that Cicero's *Dream* was afforded its adventurous independent life apart from its parent text (indeed, the rest of the *Republic* itself is still incompletely recovered to this day).

5. See, e.g., McCluskey 1998; Eastwood 2002.

This means that the *Dream,* if we are to see it in its broadest historical context, presents us with a kind of doubling and shifting where the Macrobius commentary acts sometimes as a useful amplifier, sometimes as a distorting mirror to themes in the main text.

At *Republic* VI.15, Cicero has the Elder Scipio (hereafter "Africanus") present a description of the cosmos that is at one and the same time both physical and political. Physical insofar as we are given the shapes and orders of the planets, and political insofar as the roles of humans in this cosmos are an integral part of the whole in two related ways. The story is told in the voice of one of Rome's great iconic heroes, the long-dead Scipio Aemilianus (hereafter "Scipio"), who had been the final destroyer of Carthage in 146 BC (about a century before Cicero wrote the *Dream*). Scipio tells us that while he had been serving as military tribune in Africa in the early days of the third Punic War, he had had a dream in which he had been visited by the ghost of his famous grandfather, Africanus (another monumental hero, from the second Punic War half a century earlier). Africanus proceeds to predict both Scipio's future glory and his ultimate death in the midst of the unprecedented political violence that came in the wake of the brothers Gracchi. But, Africanus bids his grandson, have no fear! Those who serve, help, and expand the republic have a place in the heavens and an eternity of happiness. So Scipio is enjoined to love justice and duty (*pietas*) to family and country alike. And at this point Scipio begins, in his dream, his iconic journey up into the heavens, a journey that would serve as a model for Dante, Milton, and even Mozart.

The journey through the heavens serves to give Africanus the opportunity to explain the layered structures and many interrelationships that unite and bind the cosmos. On the one hand, we are told that human souls are "from" the stars (*ex illis sempiternis ignibus*). This material connection underpins the set of astrological relationships that Cicero points to as Scipio's gaze is directed from the earth up into the heavenly realm.

> quaeso, inquit Africanus, quousque humi defixa tua mens erit? nonne aspicis quae in templa veneris? novem tibi orbibus, vel potius globis, conexa sunt omnia, quorum unus est caelestis, extimus, qui reliquos omnes complectitur, summus ipse deus arcens et continens ceteros, in quo sunt infixi illi qui volvuntur stellarum cursus sempiterni; cui subiecti sunt septem qui versantur retro, contrario motu atque caelum.[6]

6. *Rep.* VI.17.

> Come now, said Africanus, how long will your attention be stuck on the ground? Don't you see what sacred spaces[7] you have come into? Everything is connected by nine spheres [or better, globes] of which one is the heavenly, the outermost, which embraces all the others. It is itself the highest god, enclosing and holding the others together, and in which those that move, the eternal courses of the stars, are fixed. Beneath it there are seven that move backward, in a reverse motion to that of heaven.

Each planet is then described in turn, and the chief astrological attributes are attached, almost epithetically, to them. Jupiter, we are told, is *hominum generi prosperus et salutaris,* favorable and wholesome to the race of men. Mars is bloody and terrible on earth, *rutilus horribilisque terris.* It is clear that these attributes are astrological, not just associations drawn incidentally from their eponymous gods. This point is most dramatically underscored by the description of the role of the sun, which is said to be the lord and ruler over the other planets, *dux et princeps et moderator luminum reliquorum.*[8] So also, the sun is the mind and balance, *mens mundi et temperatio,* of the world. *Mens mundi* and *princeps* are standard astrological homonyms for the sun, and we see that Cicero's wording is reproduced almost verbatim four centuries later by the astrologer Firmicus Maternus, who calls the sun *mens mundi atque temperies, dux omnium atque princeps.*[9]

This brings us to the second respect in which humans are integral to the cosmos: for Cicero there is an important idea that it is *mens,* mind, whose proper job it is to rule, and this holds for both the cosmic and mundane levels. Not only does the sun, the mind of the world, rule the cosmos, but it is also humanity's role to watch over (*tueor*) the earth, where that position is described in legal terms as coming about by a law. It is also Scipio's *mens* that the elder Africanus is repeatedly trying to draw up from the earth to look on the heavenly sights when he becomes repeatedly fascinated with looking down at his earthly home. Cicero associates the role of human *mens* with the fact that humans have been given soul, where mind and soul are explicitly related in the next clause with reference to the stars, which are said to be *divinis animatae mentibus.* Finally, Scipio is told that his true self is not his body, but his *mens,* and it is at this point that the astrological

7. *Templa* is a loaded term in this context, to which we shall return shortly.

8. Compare Ptolemy, *Tetr.* I.2.3, where he uses the verb κατακρατέω; Manilius, I.19, opens his poem by praying that his verses will be *Phoebo modulante;* and Sextus Empiricus, *Adv. math.* V.31–32, where the sun is called βασιλεύς.

9. Firm. Mat., *Math.* I.10.14.

motif gets its fullest flowering. Where the sun was the *mens mundi* and the *dux et princeps et moderator* to the other planets, Africanus now gives divinity to human *mens* on the grounds that it governs (*moderatur*) the body just as god governs the world.[10]

> deum te igitur scito esse, si quidem est deus qui viget, qui sentit, qui meminit, qui providet, qui tam regit et moderatur et movet id corpus cui praepositus est, quam hunc mundum ille princeps deus; et ut mundum ex quadam parte mortalem, ipse deus aeternus, sic fragile corpus animus sempiternus movet.[11]

> Know then that you are a god, if indeed a god is that which lives, which perceives, which remembers, which has foresight, and which rules and governs and moves the body over which it is set, just as the supreme god does the cosmos. And just as the immortal god moves this world, part of which is mortal, so our eternal soul moves our fragile body.

We now have a three-tiered symmetry of divine rulership suspending from the idea that *deus/mens* is *dux, princeps,* and *moderator*: a divine and eternal god rules the cosmos, the divine and eternal sun rules the other planets, and the divine and eternal human soul rules the human body. Governance, reason, mind, humanity, and cosmos all come together under one set of relationships and responsibilities. The physics is at the same time politics, psychology at the same time ethics and duty. The planets, as astrological actors, mediate as both a model and as a physical causal mechanism between the (paradigmatic) order of the heavens, and the rational system of state down on earth. (Here, the *Dream's* place at the end of Cicero's great political treatise, the *Republic,* is worth keeping in mind.)[12]

The physical description of the world that interleaves with the political one parallels the earth with the cosmos as a whole in still more ways. Cicero describes both the stars themselves and their courses with words meaning "sphere" or "circle": *globus, orbis, circulus, rotundus,* which match his descriptions of the earth as a *globus.* Indeed, if we are to be precise, his description of the order of the planets is a description not exactly of the planets but of circles of planetary movement (recall the *qui volvuntur, stellarum cursus sempiterni* at VI.17), each of which is said to *belong to* a planet: *unum*

10. *Rep.* VI.24.
11. *Rep.* VI.26.
12. For the integration of the Dream and the dialogue as a whole, see Gallagher 2001.

globum possidet illa (sc. stella), quam in terris Saturniam nominant.[13] Descriptions of the remaining planets are structured in parallel: *deinde . . . ille (sc. globus) . . . qui dicitur Iovis, tum . . . quem Martium dicitis, deinde . . .* and so forth. The description progresses down to the ninth sphere, which is just the earth, immoveable and at the center of the cosmos. In fact, this is quite a remarkable description, insofar as the earth is not usually included as just another sphere in the grand scheme. In general, ancient cosmological descriptions, although they acknowledge the sphericity of the earth, do not situate it as one of the cosmic circles. Instead, the usual descriptions talk about the seven planets plus the sphere of the fixed stars, and break off the narrative with the moon. (Then again, most descriptions are looking from the ground up, not from the top down.) Cicero acknowledges the traditional point of disjunction at the lunar sphere insofar as he says that what is below the moon is mortal, that the earth itself does not move, and that it has a special relationship with ponderous bodies (they fall toward it). But he simultaneously co-integrates the earth and the cosmos in a new way.

All of this gets tied together by Cicero with a rich, if subtle, analogy. He repeatedly calls the cosmos a *templum,* which metaphor points up and down simultaneously: a *templum* is not only an earthly shrine to the gods, but—and this fact would be fresh in Cicero's mind, since he was appointed to the college of augurs in the middle of writing the *Republic*—*templum* is also the technical name for the sacred observational space marked out in the sky by an augur, the very window opened between the gods and men where they communicate.[14] It is also a *templum* that the soul of Romulus goes into during an eclipse,[15] an entrance that refers ambiguously to one or all of the following events: the conception, birth, or death of Romulus, or else his founding of Rome, all of which were associated with eclipses in antiquity.[16] We see again a multilayered paralleling of the physical and political, spatial and religious, past and future. Moreover, the various parallels serve to reinforce each other. As we saw earlier, it is not just that the order of the cosmos underpins arguments for moral (in this case political) order, but also that the moral order underpins the cosmic—precisely in the

13. Compare Pliny, *NH* II.

14. Cicero was appointed as augur in August of 53. He had begun writing the *Republic* in May of 54 and had completed two books by October of that year. We know the whole was finished by 51. On *templum* in its many senses, see Linderski 1986, 335f.

15. *Rep.* VI.xxii.

16. To hold that all four events were associated with eclipses is astronomically impossible, unless Romulus were born exceptionally prematurely. What we have instead are two distinct traditions, one in Plutarch, *Rom* 12.6 and one in Cassius Dio, I.5.12. The first has eclipses at Romulus' conception and the founding of Rome, the other at his birth and death.

divinity and rationality of the cosmos itself. Again, we find ourselves on a two-way street.

Robert Gallagher has further shown that there is a far-reaching and complex set of metaphors at play in the terminology that Cicero uses to describe the cosmos in the *Dream*, and the terminology he uses to describe political changes in the *Republic* as a whole.[17] Words such as *cursus, conversio, orbis,* and others are used to flag fundamental political processes in the *Republic,* and astronomical ones in the *Dream.* Moreover, the use of astronomy as a framing device for the whole dialogue is shown by Gallagher to be playing a central role in highlighting important aspects of Cicero's political argument (the work begins with a description of a wondrous astronomical device built by Archimedes and ends with the *Dream*). Nevertheless, there is more than just metaphor at work. In the *Dream,* we are told twice that nothing is more pleasing to god than political duty. The cosmologization of political movements serves not as an analogy, but as ground and explanation for man's moral duty in political action. Which is odd, because on the face of it the *Dream* seems to be downplaying human achievement insofar as the smallness of the globe and of the empire is an important lesson offered by the celestial perspective of the heavenly journey of Scipio and Africanus. We might object that it is small-minded *ambition* that is the real object of scorn, but we would still be faced with the tension that, on the one hand, there is a special place in the sky for anyone who has preserved, helped, or expanded the republic,[18] while on the other, Scipio is repulsed by the insignificant extent of Roman power when seen from a great height, *quo quasi punctum eius attingimus,*[19] and Africanus later reemphasizes its insignificant size, *qui tamen tanto nomine quam sit parvus.*[20]

While it is true that there is a remarkable repetition of terminology from one realm to the other in the *Republic* as a whole, it is not clear precisely what the "revolving" (*convertere*) from, say, kingship to tyranny under Tarquinius Superbus has in common with the revolution (*conversio*) of the stars.[21] As Gallagher handles it, the fundamental point is this: that there is a rational intelligence behind the motions of the heavens, just as there is a fundamental role for *mens* in the establishment of governments. The idea is further underscored by the special place that right politics plays in god's

17. Gallagher 2001.

18. *Rep.* VI.13. One attempt to clarify the apparent contradiction has been made by Powell 1996.

19. *Rep.* VI.16.

20. *Rep.* VI.21.

21. *Rep.* II.44–7, pointed to in Gallagher 2001, 511.

heart, where god loves nothing more than seeing good citizens, motivated by all the right factors, acting to benefit the republic. The point is ethical rather than historical: political good is cosmic good when properly understood.

But there is another aspect to this that lies in Cicero's use not of astronomy, but of astrology. We have already seen that the planets have particular astrological characteristics in the dream. There is another important astrological phenomenon discussed by Cicero: the idea of the *Great Year.* This is the cycle in which all the stars and planets return to the same configuration once again after many, many years, and the entire sequence of cosmic motions begins anew. The idea is rooted in the recognition that each planet has a speed of its own, and that the cycles of any two of the planets will harmonize with each other, at least very nearly, in some calculable time span. It was known, for example, that lunar months have a nineteen-year cycle, whereby a new moon observed on, say, the vernal equinox will repeat on the same date in very nearly exactly nineteen years' time.[22] This is because certain cycles of the sun and moon very nearly intersect every nineteen years. Something analogous happens with each of the other planets as well, though the numbers are reckoned differently in different sources.[23] If we wish to take the periodicity of more than two planets into account, we start having to multiply their individual cycles relative to each other, and the combined cycle after which all the phenomena start to repeat accordingly gets larger and larger the more bodies it is trying to account for. Thus the calculation for the period in which the cycles of all the planets together collectively repeats becomes a very large number of years, sometimes reckoned in antiquity at 1.75 million, but on at least one occasion calculated with more exacting standards to roughly 648 quadrillion years (by taking longer values for each planet's periodicity, every improvement in accuracy multiplies through to give a longer value for a Great Year). Whether Cicero had even an order of magnitude in mind, let alone an exact figure, is not known. Nevertheless, the idea of the phenomenon was an important one in ancient astrological contexts and held a powerful grip on the ancient imagination.[24]

Cicero's mention of the Great Year is doubly interesting. He offers it as a way of doing to human time scales what the rise into the heavens did for

22. By "date" here I mean seasonal date. After the Julian calendar reform was implemented and corrected, seasonal dates could then be equated with calendar dates.

23. See Neugebauer 1975, 606f.

24. This is not to impute to Scipio an assertion of the eternal recurrence, but rather to point out the astrological reverberations that any reference to the Great Year would have had in Cicero's audience.

human geographic scales: you thought the empire was big, but it is actually quite small; you thought that generations of Romans would sing your praises for many ages to come, but in fact your fame will be but a blip in cosmic time. Cicero also drops the Great Year into the discussion in the context of Scipio's mention of the Stoic conflagration and the recurrence of the same that the conflagration implies.[25] The Great Year and the recurrence of the same underscore each other partly just implicitly, but also explicitly insofar as they were generally taken as part of the same package in the Stoicism of Cicero's day.

Again, the astrology stands as the middle term in a three-tiered symmetry where the roles of god are paralleled by those of the planets and by our roles on earth. Here it is whole realms that mirror each other, even as particular entities manifest in each of the three realms (e.g., *mens* and *animus*), and the whole thing is wrapped up in a package that repeatedly plays on ideas of cyclicality. Moreover, as the astronomy gets fleshed out in the dream, the cyclicality of heavenly motion does more than just produce the Great Year. It also produces an especially beautiful music.

Numbers in Nature

To most ancients, it was no coincidence that the number of planets equaled the number of notes in the musical scale. Ideas of attunement and harmony were likewise easily applicable to human bodies, either directly (as harmonious sound) or via an astrological intermediary.[26] Much to the chagrin of some of his modern commentators, Ptolemy ends his *Harmonics* by using his mathematization of musical tone and interval to analyze and explain a host of properties of the human soul, including the idea of virtue and our emotional states, as well as geometry and cosmology more generally. All this is happening as the culmination of a work often characterized as one of Ptolemy's—indeed one of antiquity's—most methodologically sophisticated, most *scientific*, texts. This symmetrical translation of harmony as an explanation from musical theory to all kinds of other realms is common. As Cicero frames it:

> quid hic, inquam, quis est qui complet aures meas tantus et tam dulcis sonus?

25. *Rep.* VI.23.

26. E.g., the Hippocratic *De diaeta* I.8; Ptolemy, *Harm.* III; Aristides Quintilianus, esp. book II.

> hic est, inquit, ille qui intervallis disiunctus imparibus, sed tamen pro rata parte ratione distinctis, impulsu et motu ipsorum orbium efficitur, et acuta cum gravibus temperans varios aequabiliter concentus efficit. nec enim silentio tanti motus incitari possunt, et natura fert ut extrema ex altera parte graviter, ex altera autem acute sonent. . . . illi autem octo cursus, in quibus eadem vis est duorum, septem efficiunt distinctos intervallis sonos, qui numerus rerum omnium fere nodus est; quod docti homines nervis imitati atque cantibus, aperuerunt sibi reditum in hunc locum[27]

> What, I said, is this sound that fills my ears, so loud and at the same time so beautiful?
>
> That, he said, is the sound, separated into unequal intervals that are nevertheless divided in perfect proportion, caused by the impulse and motion of the stars themselves, and the balanced intermingling of highs and lows causes the varied harmonies. Such great motions could not be brought about silently, and Nature makes it such that the extremes make sounds, low at one end and high at the other. . . . These eight circuits [of the planets], two of which have the same power,[28] produce—and this number is like the node of all things—*seven* different sounds in intervals. Thus learned men, imitating these with strings and with their voices, have laid open a return for themselves to this region.

Notice that the *docti homines* are said to *return* to the heavens rather than just travel up to it. Here the soul's journey from the heavens into the body and back to the heavens is seen as a cycle deliberately paralleled with the sun's annual return, which cycle is repeatedly called a *reditus* in the *Dream*, just as the return to the stars is here.[29] Earthly music imitates the sounds of the heavens. But it also produces a smaller version of the cycle of life: where virtuous men's souls return to the sky after death, so they temporarily travel to the heavens in the contemplation of earthly music that imitates that of the sky. Via a Heraclitean allusion, *reditus* serves to doubly under-

27. *Rep.* VI.18.

28. Ptolemy's *Canobic Inscription* (published by Heiberg in vol. 2 of Ptolemy's *Opera*) explicitly tells us that Venus and Mercury make one and the same note. Macrobius, like many commentators since, treats the *vis* in terms of planetary (mean) speed, and so he equates the speeds of Mercury and Venus (a point that goes back to Plato, *Tim.* 38c). But it is significant that Plato does not just equate the speeds of Mercury and Venus, he equates them *both* with the sun, which would mean *three* spheres have the same speed, as he says explicitly at *Tim.* 38d, not just two.

29. E.g., *Rep.* VI.22, VI.12.

score the way that music itself is produced, as well as its heavenly origins.[30] As Heraclitus said:

> παλίντροπος ἁρμονίη ὅκωσπερ τόξου καὶ λύρης.[31]
>
> Harmony is a back-turning, as in a bow or a lyre.

Here harmony is the product of tension, of forces opposing by turning back on themselves. ἁρμονία is an obvious reference to musical harmony in this passage, but its primary meaning in Greek is simultaneously linked to Cicero's *nodus,* a binding or joining. There is also a fairly rich set of astronomical metaphors at play in the space between Cicero and Heraclitus. Both the lyre and the archer are important seasonal constellations, one marking the beginning of autumn, and the other marking the winter solstice (τρόπος in Greek, *nodus* in Latin).[32] Cicero's *nodus* thus does a clever double duty, pointing to both Heraclitus' ἁρμονία and the τρόπος that characterizes it in παλίντροπος ἁρμονίη. Cicero then frames the whole with the idea of a *reditus nervis,* a return by means of strings, as in a lyre.[33]

But it is also a *reditus,* this time of the sun, that flags the other mention of that *nodus* of all things, the number seven, in the *Dream.* The ghost of Africanus dates a prophecy about Scipio to a time *cum aetas tua septenos octiens solis anfractus reditusque converterit,* "when your life looks back on seven times eight circuits and returns of the sun."[34] Again, this passage plays on imagery of turning and revolving, circling and return. It is this instance of the number seven that Macrobius latches onto so dramatically, farming Cicero's one sentence for what is nearly twenty-one pages of the Teubner edition of the Latin *Commentary.* If anything in the *Commentary* can stand for Macrobian exegetical excitability, this is it.[35]

30. That Cicero was familiar with Heraclitus is clear from passages in, e.g., *De div.* (II.133) and *De fato* (17), although to be sure he nowhere quotes the particular line I cite here.

31. Heraclitus DK 51 (Hippolytus, *Ref.* 9.9.4).

32. See, e.g., Pliny, *HN* XVIII.220f.

33. Recall also that of all the philosophers listed in Macrobius' doxography of the soul, it is Heraclitus who comes closest to Cicero's point about human souls coming from the stars. See *Comm.* I.14.19.

34. *Rep.* VI.12

35. Stahl 1962, 95, comments that in this passage "it becomes evident that Macrobius' work is really not a commentary on *Scipio's Dream.*" By contrast, Armisen-Marchetti 2001, has a very interesting and balanced reading that dissolves the question by situating the criteria for its solution within Macrobius' Neoplatonism itself. It is still unfortunately common for even those who have tried to give Macrobius some credit for not completely perverting Cicero to be quick to distance themselves from the number-theoretical aspects of Macrobius. In discuss-

The number seven reverberates for Macrobius everywhere. It is one of the most important measures in the cosmos, and one that repeats across as many different realms as possible, from physics to physiology, from tides to teething. It is seven that forms one of the important nodes of cosmic symmetry in the *Commentary*, and Macrobius occasionally seems apologetic that he must treat it so summarily (a *mere* twenty-one pages!). We find seven to be central to number theory, to geometry, to the natures of bodies occupying what we would call space, and to the Platonic account of creation. We find it, directly or indirectly, in the heavens (in the moon, in the sun, in the planets), in the elements, in geographical zones, in the seasons, in the tides, in musical harmony, in Homer, in the World-Soul, in the ocean, in human development (in conception, in gestation, in sex determination, in growth, in teething, in speaking, in the ages of maturation), in menstruation, in the organs of the body, in medicine, and more. (Obviously not all of this would or even could have been in Cicero's mind when he parsed Scipio's age, but that is beside the point for the moment.) What is clear, is that Macrobius sees the world quite immediately as being seven-rich, and *sevenness* is an important property that relates a wide range of phenomena.

It is not just the range, though, but also the hierarchy that is interesting. While many of his examples seem odd or even trite (the ages at which children get their teeth, for example) some are absolutely fundamental to the function and being of the cosmos as a whole (the World Soul, physical space, creation, the heavens). Seven does not just turn up frequently, it turns up significantly, and its importance and extension condition the very eyes Macrobius has for the world.

In parsing the relationship between Cicero's cosmos and that of Macrobius, it is worth remarking that Cicero does, after all, call special attention to the numbers seven and eight by flagging them explicitly as *full,* a reference that carries rather a lot of baggage in its train, and he explicitly puts seven on a fairly high pedestal as the node of all things. It is also the case that, like Macrobius', Cicero's own cosmos as painted in the *Dream* is a richly intersecting place. Just by tagging and tracking *reditus* and its related ideas in the *Dream,* for example, we saw reflections of the political in astronomy, of astronomy in temporal cycles (annual and cosmic), of temporal cycles in theological themes (the soul's relation to the stars and to god), musical scales, and more. God, the stars, time, politics, and humanity all

---

ing Macrobius, even Armisen-Marchetti, for example, throws the word "arithmetic" into scare quotes to flag that she is really talking about "une mystique des nombres" rather than real mathematics (Armisen-Marchetti 2001, xlii). Compare also Powell 1990, 131–32.

interrelate and move through each other. Roles and relationships happen in similar ways across different realms, and the universe is highly symmetrical. This symmetry serves to guide Cicero's thinking about the world and his place in it. But it also allows Cicero to draw inferential conclusions in one realm based on relationships in another. If mind works one way in the heavens, then it should work in an analogous way on earth, for example. The particular points of reflection and intersection that he chooses deeply affect the way the world itself falls in line for him, simply because they play formative roles in how he is organizing his thought about the cosmos as a whole. True, he is working on a more concrete set of examples and to a much more practical end than Macrobius, but he is doing something analogous in his overall project. The second point is that it does not seem possible to call one of these levels of analysis the fundamental one. We do not have a beginning upon which the rest is grounded. Nature does not underpin politics exclusively; they underpin each other reciprocally. The web this creates is all the stronger for being grounded at multiple points and reinforced along multiple axes.

In explanations, breadth of extension is often seen as a virtue. Taking it too far, though, we run the risk of broadening explanations to the point of vagueness, and vagueness is clearly an epistemic vice. Discussion of what breadth means, and of when and to what extent it is an attractive feature of explanations, has been taken up in several modern disciplines, most notably philosophy and cognitive psychology.[36] The basic idea is just that some hypotheses or theories explain more things than their competitors do. By its proponents, it is often argued (or assumed) that a broader explanation is or should be preferred (*ceteris paribus*) to a narrower one.[37] So, to take a classic example, we should have preferred oxygen over phlogiston in 1795, since oxygen could explain all the phenomena phlogiston could, plus it could also explain weight and air-volume changes in combustion and calcination. (It is also often remarked that Newton's gravitational inverse-square attraction

36. Terminology varies a little by discipline and subdiscipline. I will follow philosophers of science and speak of breadth, but we also find discussions of explanatory versatility and explanatory range.

37. There has been considerable debate on whether explanatory breadth (sometimes handled under the auspices of accommodation or retrodiction) should be less important than the prediction of novel phenomena in the epistemology of the sciences. Lakatos 1970 is the fulcrum for many arguments favoring prediction (see also, e.g., Giere 1983; Maher 1988; White 2003). Brush 1989 and 1994 argued that in practice, scientists do not show a marked preference for prediction over retrodiction. Others have argued against an epistemological preference for prediction over breadth or retrodiction (see, e.g., Laudan 1981a; Collins 1994; Achinstein 1994). I still think Brush and Laudan's are the best set of arguments in this debate.

explains falling bodies in addition to explaining the trajectories of ballistics and orbiting planets, whereas Aristotelian natural motion only explains falling bodies—but this forgets that natural motion *also* explains planetary orbits and ballistics; we may want to make an appeal to elegance or simplicity for Newton, but breadth alone will not do.)

In parsing the use of seven as an explanation in the *Dream*, we are faced with the problem that Cicero does not explicitly line up all the ducks that would allow us to explicate the mechanism of sevens without turning to other sources for possible models.[38] Nevertheless it seems clear that, if pushed, Cicero thought he *could* have given a coherent and complete causal account, and there are certainly several potential candidates available, even if he does not exhaustively document all the links in the causal chain for his reader. On the one hand, we need to be very careful about reconstructing the details. On the other hand, it is just in the explanatory gaps, in what Cicero expects his reader to take for granted, that we can see some of the deepest-rooted assumptions that hold the whole edifice together. We may not know exactly what it is, but we can see that something is there, something big, and something already both causally integrated and thoroughly mathematized.

Harmony and Empiricism

Broad explanations are clearly symmetrical (because their identical analytic structures are applicable across diverse material or at diverse levels), and symmetry—as I've parsed it in this chapter—is clearly meant to have explanatory power, even when causal chains are not fully articulated. Nevertheless, even those modern readers who support the idea that broad explanations are in general preferable to less broad ones may be forgiven for a moment for thinking that the very high symmetry in Cicero, Ptolemy, or Macrobius represents a system that has gone wrong somehow, a system that is so very broad as to be simply vague verging on empirically meaningless, and one that seems indicative of some degree of inherent uncriticality or overinterpretation on the part of our ancient authors.[39] (Thus even one of the

38. We do sometimes find strong hints that numbers in themselves have some kind of direct causal effect, as when Macrobius says that the first two cubes—8 and 27—added together and then multiplied by 6, interfere with the normal course of human gestation to produce a seven-month baby. Macrobius, *Comm.* I.6.14. In general, though, much of the causal detail of the explanation is left unstated or understated. On "numerology" and the seven-month baby, see Hanson 1987.

39. This is the substance of one part of Aristotle's objection to the Pythagoreans at the end of the Metaphysics.

most careful and insightful of modern readers of ancient scientific method is tempted to call such broad conclusions "metaphysical and mystical.")[40]

Putting the symmetrical uses of number in their larger context, however, one where the ordering relationships (differently: laws) that govern both humans and the cosmos as a whole are so very different, where an active, divine reason is instantiated in the very material of the universe and its interrelations, we can begin to see how the investigation of numerical symmetries is instead an exercise of the highest human faculty in pursuit of the highest and most complete kinds of knowledge. Theology, ethics, politics, and law have shown their deeply ingrained relationships to the sciences again and again in this study—and it should occasion no surprise to see the broadest explanations, the grand unifying theories, reaching across all of them. The ways in which such explanations will be expected to work, the joints at which they will cut or the paths along which they will run, are exactly those most maximally embedded or entangled in the webs of ancient knowledge. Seeing Cicero's or Macrobius' explanatory use of number as primarily an exercise in mysticism or numerology loses sight of two things: (1) the important explanatory functions of symmetry (including numerical symmetry), and (2) the observational grounding of the numbers so used.

For Cicero, we can look back to Plato's *Timaeus,* the *Myth of Er,* and to Platonic rereadings of the Pythagorean tradition as sources of his ideas on cosmic harmony and the interrelations between the human soul and the planets via music theory, although I would be very careful about how we reconstruct the details.[41] Developments in the technical discipline of harmonic theory notwithstanding, the idea that the laws governing many diverse kinds of natural regularities were in some way harmonic seems to have been accepted by a wide range of thinkers by Cicero's day, not just Platonists and the successors of Neopythagoreanism. Thus we find Balbus the Stoic, in Cicero's *De natura deorum,* claiming that everything from the seasonal flowering of plants to the regularity of the tides to the orderly motion of the planets is an instance of an overarching musical harmony (*concinens*) governing natural processes under an all-pervading divine rationality.[42] Even if we did suppose that, *Zeitgeist* aside, Cicero was harking specifically back to Plato to find causal links between numbers, cosmos, and psyche, we are faced with a very difficult hermeneutic problem, the question of exactly how Cicero understands Plato's physics. We have insights into small parts

40. Barker 1989, 7

41. On how Platonists reworked Pythagoras, see esp. Burkert 1962; Kahn 2001.

42. Cicero, *De nat. deor.* II.19.

of Cicero's interpretation scattered throughout his works; we have some of his translation into Latin of the *Timaeus;* but the details are just enough to indicate how different Cicero's physics is from Plato's without really giving us a way forward. To take just one instance, we hear Cicero in the *Academica* calling Stoicism, including its physics, a "correction" of the old Academy, rather than a truly new system: *verum esse autem arbitror, ut Antiocho nostro familiari placebat, correctionem veteris Academiae potius quam novam aliquam disciplinam putandam.*[43] This is not to say that Cicero is not sometimes also a sharp critic of certain Stoic doctrines, but it highlights the fact that his Platonism is extraordinarily flexible.

A full and detailed causal discussion of cosmic harmonies, one that goes into more detail than we find in Cicero, Plato, or Nicomachus, and one in which both a physics and a psychology are worked out explicitly, is in fact not available until Ptolemy, although we do find a sketch in book 3 of Aristides Quintilianus, who may predate Ptolemy by a little. No one who reads Ptolemy's *Harmonics* can fail to be impressed by it. In it Ptolemy does all manner of things that historians of science used to think no one had done before the Renaissance. He performs very careful tests, isolates individual variables, corrects for errors and limitations in his instruments and in the human sensory apparatus, and he experiments.[44] The mathematics he employs is very sophisticated and the book is exceptionally difficult and highly specialized. One might try to appeal to these hurdles as explanations for why the *Harmonics* is so seldom read in the history of the sciences—but then the same objections of difficulty and obscurity would apply to Ptolemy's much longer *Almagest,* which though not frequently read, is still at least part of the canon of the history of the sciences. This is unfortunate, not only because the *Harmonics* is such an interesting and rewarding work, but because it is indicative of a more general ignorance of how important the idea of harmonics was to ways of approaching nature, both before and after Ptolemy.

For Ptolemy the starting point is the realization that harmony, as a kind of motion, will have particular kinds of relationships with matter and form. Moreover, harmony is not just any random motion, but is instead *rational* motion. This means, then, that it is best found in those kinds of matter that are themselves most rational. That harmony manifests both rationally and materially at the same time means that the investigator into harmony must employ both reason and a carefully trained and constrained empiri-

43. Cicero, *Acad.* I.43.
44. See Barker 2000.

cism to uncover it. This trained rational empiricism, says Ptolemy, is what the science of "mathematics," taken most generally, is really all about.[45] In its purest forms, it employs the two senses—hearing and sight—that are located physically closest to the *hegemonikon,* the governing faculty of the human soul. That these really are the two epistemologically primary senses is further shown by the fact that, whereas the other senses can judge basic pleasantness and unpleasantness, only sight and hearing convey beauty (τὸ καλόν). The interrelationship between sight, hearing, and reason runs deep: τὰ μὲν ὁρατὰ μόνως ἡ ἀκοὴ δεικνύουσα διὰ τῶν ἑρμηνειῶν, τὰ δ' ἀκουστὰ μόνως ἡ ὄψις ἀπαγγέλλουσα διὰ τῶν ὑπογραφῶν,[46] "Only hearing brings to light things seen (by means of descriptions), and only vision brings back reports of things heard (by means of writings)." The idea is pretty, the inversion in the turn of phrase very clever: the eyes "bringing back reports" and the ears "showing." Moreover, Ptolemy is here underscoring the interpersonal and communicative nature of knowledge—it is a project that requires both conversation and reading—including a recognized dependence on testimony. Finally, the soul itself functions best when the two senses are used in concert, as when diagrams accompany explanations, or poets describe visual scenes.

Each of these two highest senses has a discipline and a subject matter specific to them, and each of their proper subject matters is governed in the same way by a skilled interrelationship between perception and mathematical reason. For vision, the highest subject matter is astronomy, since astronomy deals with the most divine of all physically observable things, the heavenly bodies. Among earthly things, the most rational of material things is the human soul, and this becomes the subject of harmonics in its purest form. The move from audible sounds to human souls happens in two ways for Ptolemy. On the one hand, there is the often remarked affinity of the human soul for music, where audible harmony is widely taken as both morally fortifying and productive of happiness. Thus Aristides Quintilianus directly relates historical periods of political harmony and disharmony at Rome to the musical cultivation of its leaders.[47] Ptolemy, too, seizes on the political and jurisprudential dimensions of harmony, likening various kinds of laws and political movements to various musical modes. This affective relationship is reinforced by mathematical analysis applied to the theory of the faculties of the soul, where various ways of dividing the soul's abilities and affectations map very closely, and with a high degree of specificity, onto

45. Ptolemy, *Harm.* III.3.
46. Ptolemy, *Harm.* III.3.94.1–2.
47. Aristides Quintilianus, II.6.

the harmonic divisions and their interrelationships. From here, it follows directly that the best condition of the soul is a harmonic balancing of the faculties and appetites, and the study of harmonics very quickly has direct ethical and political consequences.

It is at this point that Ptolemy turns his eye upward to the heavens, arguing that astronomy "finds its completion" by means of the harmonic ratios (τὰς τῶν οὐρανίων ὑποθέσεις κατὰ τοὺς ἁρμονικοὺς συντελουμένας λόγους).[48] In the first instance, heavenly motion is of the same species as musical harmony, in being fundamentally determined by intervals (διαστηματική) and in being inherently cyclical and highly ordered. This is all well and good from a theoretical point of view, but Ptolemy now does something remarkable. He makes an empirical turn. He begins by imagining the standard harmonic unit, the double octave, as a line, and then he superimposes it on the heavenly grid of the ecliptic, with the first and last notes in the double octave situated at one equinox, and the middle point of the double octave situated at the other equinox. Ptolemy then begins to plot out the most significant of the astrological influences between zodiacal signs, the relationships called trine, square, and opposition, where a planet in one sign can be affected very strongly by another planet in another sign when the two signs stand in given geometrical relationships to each other (fig. 8.1).

Counting the number of signs anticlockwise from Virgo round to Taurus, its trine, we find that Taurus stands eight signs away. From Virgo to Gemini, its square, we find nine signs. Virgo to Pisces (opposition) is six. Going clockwise, we find Virgo to Taurus is four signs, Virgo to Gemini three. Now the most important of the harmonic ratios are the octave (2:1), the fifth (3:2) and the fourth (4:3). In harmony, these ratios are known to have the most powerful effects on the human soul, just as in astrology trines, squares, and oppositions are known to be the most powerful relationships. Ptolemy now shows us why the latter should be the case. Quite simply, by counting signs from trine to square, or from square to opposition, or from opposition to trine, he is able to show that all the significant signs stand to each other in relations identical to the primary harmonic ratios. Thus from Virgo to Taurus and Virgo to Gemini (trine and square) clockwise, we see a harmonic fourth (4:3), and from Virgo to Taurus and Virgo to Pisces (trine and opposition) counterclockwise, we find a fifth (3:2). As he explores the various positions of the signs to each other, Ptolemy finds three different instantiations of each of the main harmonic ratios (three octaves,

48. Ptolemy, *Harm.* III.3.100.24–26. The difficult grammatical construction is interestingly repeated here from 94.10.

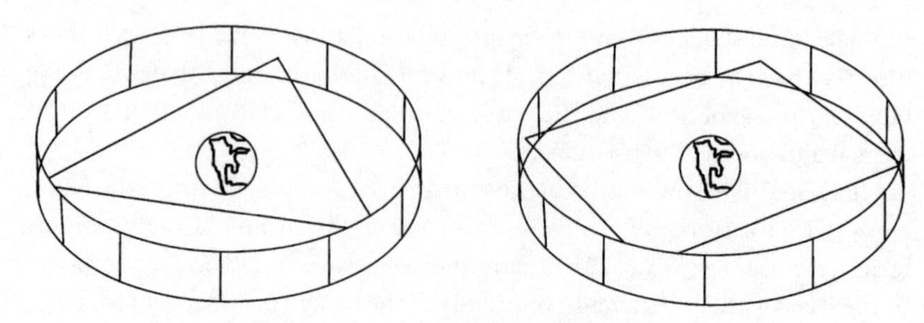

Fig. 8.1 Trine and square zodiacal signs

three fifths, and three fourths) just in trine, square, and opposition alone. Other concords are likewise present, with two instantiations of the octave-and-a-fifth, one double octave, one octave-and-a-fourth, and even one of the humble single tone. Ptolemy further shows that signs standing in nonconcordant ratios to each other also consistently stand in uninfluential astrological positions and vice versa. Finally, if we now imagine the sun in some position on our zodiacal circle, and think about what positions the moon (or any other planet) has relative to the sun at its significant phases—full moon, new moon, first and last quarter, and so on—we find not just familiar concords, but also very specific alignments corresponding to the notes that divide the double octave in Greek music.

One might worry that if humans had divided up the zodiac differently (into, say, thirteen signs instead of twelve), the numbers would not work out so well. For Ptolemy this is a nonstarter. The twelveness of the signs is explicitly said by Ptolemy to have been established *by Nature,* δωδεκαμερῆ τὸν τῶν ζῳδίων κύκλον συνεστήσατο ἡ φύσις,[49] and we know the number through observation of each sign's discrete effects.

Conclusion

In reading book 3 of the *Harmonics,* if one follows along in the careful commentaries by Barker and Solomon,[50] one sees again and again that Ptolemy borrows ideas heavily from many different thinkers and schools. One distinction is Stoic, another definition is Aristotelian, a third idea Platonic (although it is worth noting that none seem to be Sceptical or Epicurean). As with Cicero, we see one group of schools being treated as compatible, their ideas interposable. However, this is not to say that Ptolemy (or Cicero for

49. Ptolemy, *Harm.* III.3.103.12–13.
50. Barker 1989; Solomon 2000.

that matter) thought Plato equaled Aristotle equaled Zeno of Citium. What it does show is that both physical problem sets and their solutions were grounded in a wide field of theoretical possibility that crossed school boundaries in many of its details. Is this eclecticism? Perhaps a better term would be "concentric," pointing more clearly to the common theoretical ground shared by the compatible schools, an agreement that begins with the idea (observation?) that the cosmos is rational and divine, and then progresses to explore the difficult problems of understanding—of knowing—that arise when experience hits the world. The question of where Ptolemy got his sevenfold division of the intellect is historically interesting to us now, to be sure, but it is paradoxically also interesting that the fact that this division was Stoic rather than Aristotelian seems not to have mattered one whit to Ptolemy himself.

The way that Ptolemy's discussion moves across the board, both in his relationships to his predecessors and in where he sees himself as contributing something new to the discussion, is telling. Even the way he so easily zooms out at the end of the work from the minute and highly controlled mathematical and experimental investigation of perceptible acoustic intervals to the broad investigation of the δύναμις καταληπτική, the perceptive function or power that is harmonics itself,[51] is indicative of how widespread the ideas are that he is building on. We cannot, of course, impute any particular one of Ptolemy's interrelationships between harmony, soul, and cosmos back onto Cicero, but we can at the same time see how rich, detailed, and above all how surprisingly well grounded the idea of cosmic harmony was when an author did bother to flesh out the details. We could similarly point to Aristides Quintilianus' discussion of cosmic harmonies in book 2, which although it is not as clearly and thoroughly argued as Ptolemy's, makes many compatible points lining up human psychology with the stars via sympathetic relations between the human body, the cosmos, and sound. Harmony was, it seems, a very good explanation for a very broad range of phenomena, uniting the highest good with the highest rationality: cosmos, god, soul, happiness, and ethical duty. What more could anyone ask of an explanation?

51. My translation of *catalēptic* as "perceptive" captures one of the important aspects of the Greek word, but not all. The word literally means "grasping," and was incorporated into the foundations of Stoic epistemology to refer to the kinds of immediate but nonformal certainty that perception, among other things, gives us. See, e.g., Frede 1999 and chapter 5, above.

CHAPTER NINE

Of Miracles and Mistaken Theories

When we first encountered garlic-magnet antipathy and its epistemological implications, I floated the Kuhnian idea of "living in different worlds" as one possible solution to the problems garlic and magnets posed. Nevertheless, I held back from espousing a full-blown relativistic reading of this idea, and I did so for reasons to be elaborated over the next two chapters. To anticipate part of the conclusion, I should say in advance that I am, in spite of appearances, a scientific realist. That is to say, I believe in the reality of electrons and phosphate bonds and a host of other entities, structures, and phenomena that I have never seen directly. I come by this realism the hard way, though, from an early commitment to historiographic relativism to a (weak) realism about atoms and genes, thanks to something like what Philip Kitcher has called the "Galilean strategy."

The Galilean strategy gets its name from a story about Galileo in the early days of the telescope. In order to overcome worries that the new stars and lunar craters he was seeing might actually be no more than artifacts of the new instrument, Galileo is supposed to have shown the sceptics that more familiar objects—distant buildings, ships at sea—were accurately magnified by the telescope, and so what he was seeing in the night sky must also be an accurate picture of what was up there. For Kitcher, the Galilean strategy for realism argues from the (given) reality of macroscopic objects like trees and dogs—if you can spray it for fleas it must be real—to the reality of unobservable objects like electrons.[1]

1. Kitcher 2001; see also Hacking 1983. Kitcher's argument is considerably more sophisticated than the short version I have presented here, and in particular he goes to some length to avoid the more obvious objections to the telescope example. Kitcher's discussion of how we understand the relationships between the schematized stations of the London underground map and their sub- and superterranian counterparts, for example, presents much more difficulty for

Having said this, though, I must admit that my realism engenders some difficulty, because a realist is in many ways a difficult thing for a historian to be. And this trouble runs both ways: realism is a problem for history, and history is a problem for realism.

The historian's problem is that too adamant a belief that the sciences basically describe the world as it is (or very nearly) can have unfortunate historiographic consequences. Thinking we now have it right, for example, has often tended to elevate certain historical theories and disciplines above others in the historian's eyes by means of an implicit value judgment that what is best in past science is what is most like modern science. So astrology and all the religious baggage of ancient and medieval science are out, and mathematical astronomy is in. Their relative evaluations at the hands of the historical actors themselves then seem to have no bearing, and we run full-speed into a vision of historical science in which present values determine past value retroactively. Anachronism becomes hard to avoid. A lovely example of this general frame of mind can be found in the Greek ten-drachma coin of the 1980s (see fig. 9.1), which had a picture of the Presocratic philosopher Democritus on one side, and the Bohr model of the atom on the other. The more egregious versions of this tendency are easy enough to spot, to be sure, but the problem is an insidious one. In this book, I have taken seriously David Bloor's insistence that explanations of why people believed in "right" science should be symmetrical with explanations of why people believed in "wrong" science. But the tendency to attribute belief in right theories to empirical evidence while attributing belief in wrong theories to psychology, religion, society, or mysticism is very difficult to avoid if one is a realist.[2] So my claim that garlic-magnet antipathy has an empirical foundation is closely tied to a radical insistence on Bloorian symmetry, including an emphasis on evidential symmetry, which position would seem to be very difficult for a realist to hold.

Starting from the other side of history's encounter with realism, we see that in philosophical debates on scientific realism, historical examples are often taken to undermine the grounds for the realist's beliefs about the world. If, the argument goes, we can show that our best sciences in the past—even those that were predictively successful—have been radically wrong about the basic entities (structures, etc.) that make up the world, then what grounds

the old objection to Galileo that "perhaps things are just different up there" (see Feyerabend 1975, e.g.). In the case of the tube map, "up there" is accessible in a way that the moons of Jupiter, in 1610, were not.

2. See, e.g., Laudan 1981b.

Fig. 9.1 Greek ten-drachma coin

does the realist have for believing that today's predictively successful science is getting things right?

Looking at the interactions of history and realism in light of our analysis of the Roman sciences, however, things get even more complex. For proponents and opponents of realism alike, there is all too often a tacit agreement that historical examples—whatever they are taken to prove—need not go back before either Fresnel's luminiferous ether, or else Lavoisier and phlogiston. This "Fresgiston Pact," this agreement that for philosophical purposes science is a two-hundred-year-old activity, has the virtue that it keeps the *really* messy unobservables—antipathy, pneuma, harmony, god—out of the debates. It seems a fitting conclusion to this book to ask what happens when we lift this barrier.

What follows is a complex argument, one that will take me the whole of two chapters to develop fully, and so I ask the reader's patience. Historians who are unsure why they should care about scientific realism at all may want to jump over the present chapter and start instead with chapter 10, where I more fully outline the historiographic implications of realism versus relativism. In the present chapter, we begin with a look at one of the standard arguments for realism, Hilary Putnam's so-called "miracle argument," which states at its most basic that either science is right about the universe, or else the success of science must be counted as miraculous. Although much criticized, the miracle argument still finds traction in contemporary debates, and we further find that some version of it has been around for a very long time. The main importance of the miracle argument for the present inquiry, though, lies in the literature it has generated both for and against, since this literature bears directly on how we understand the content of the historical sciences. Premodern sciences, it turns out, pose difficult problems for the miracle argument, since the entities and processes that the ancients thought they were right about—and that they thought they were successfully manipulating—turn out not to exist at all. In light of this, philosophers have tried to save the miracle argument by demarcating a set of criteria according to which we can say that the modern sciences are importantly and sufficiently different from their premodern counterparts, so that we can now legitimately have confidence in the entities (structures, processes) that our science posits, whereas our forebears were wrong in drawing their parallel conclusions. There are two different strategies generally (and often overlappingly) employed here: one says that there is something characteristically different about modern science that makes its successes more reality-indicative, and the other says that we can access what-is-real if and only if we are very careful about which aspects of our theories we take to be reality-indicative, and that our forebears put their faith in the wrong aspects.

I will argue that, in fact, both of these strategies are doomed to fail, the one because it commits a logical fallacy (*ad novitatem*), and the other because it is impossible to show that its criteria are sufficient. This puts the realist in the awkward position of having to accept both a changing and a changeable content for the box of things labeled "real." It does not, however, necessitate a blanket move to relativism, as I will argue in the next chapter. The trick lies in carefully handling the point of contact between realism and what has often appeared to be its natural bedfellow: the correspondence theory of truth (this is roughly the theory of truth that Aristotle proposed when he said "'true' is to say of what-is *that it is* and of what-is-not *that it is not*").[3] This necessitates placing a stronger emphasis on *justification* than is normally done, which will further require a move away from foundationalist epistemologies (here I follow Hasok Chang's lead) and toward a pragmatist approach to justification, truth, and meaning.[4]

We begin with a look at the so-called miracle argument for realism, not because it is itself a point of departure for my argument, but because the challenges that have been arrayed against it and the realist attempts to block those challenges bear directly on any historically informed discussion of realism. In the end, we will see that the standard moves that realists use to cordon off the historical objections to the miracle argument (appeals to

3. Aristotle, *Metaphysics* 1011b25. Italics mine.
4. Chang 2004.

maturity and predictive success, and what I call "curate's-egg" approaches) all rely on the ungrounded assumption that we—now—occupy an epistemically privileged historical position, one that allows us to pass judgment not only on the realisms of those who went before us, but on our own as well, and that further assumes (falsely) that the criteria for this judgment can be transhistorical.

History as a Problem for Realism

This is Putnam's original statement of the miracle argument:

> The positive argument for realism is that it is the only philosophy that doesn't make the success of science a miracle.[5]

Compelling, on the face of it; so compelling that it is not infrequently characterized as "the ultimate argument for realism."[6] Nevertheless, the miracle argument is hardly ironclad.[7] Its power is indicated by the way that it stands at the heart of what we might call the realist intuition. More sophisticated versions of realism—and there are many—do not begin from the miracle argument, but again and again, arguments for realism still try to make sense of the important connection that stands between predictive success and truth in the miracle argument. For all that the miracle argument is not self-sufficient, the connection between our high degrees of success and the truth about the world is still central to realism, and most versions of realism go to some lengths to show why that connection persists. It is, to paraphrase a little, the ultimate intuition for realism, and it needs accounting for in most any version of realism one wants to name.

The historiographic line I have been arguing in this book should give us significant reason to pause at this point. My caveat—not yet an objection, but a red flag all the same—is this: some version of the miracle argument was tenable at almost any point you want to name in the history of the sciences.[8]

5. Putnam 1975, 73. More fully fleshed out in Putnam 1978, 18f.

6. E.g., Musgrave 1988.

7. Chakravartty 2007, chap. 1, has a nice summary of problems with it. The seminal challenge is Laudan 1981a.

8. I readily admit that this claim is not exceptionless, but to make my case the one instance suffices. Kitcher has also mentioned (although drawing a different conclusion from mine) the role of the miracle argument in Descartes' realism about his own theories (Kitcher 2001, 166). Nick Denyer was also correct in drawing attention to the similarities between Quintus' argument for divination and the miracle argument (Denyer 1985). Other examples abound.

If it motivates realism about DNA and electrons now, the miracle argument should have motivated realism about sympathy and antipathy, luminiferous ether, caloric, and pneuma at various points in the past. Indeed, we find something very like the miracle argument at the very beginnings of classical science, in the Hippocratic work *On Ancient Medicine*:

> διὰ τὸ ἐγγὺς οἶμαι τοῦ ἀτρεκεστάτου δύνασθαι ἥκειν λογισμῷ ἐκ πολλῆς ἀγνωσίης θαυμάζειν τὰ ἐξευρημένα, ὡς καλῶς καὶ ὀρθῶς ἐξεύρηται καὶ οὐκ ἀπὸ τύχης.
>
> Because of the near perfect accuracy at which [medicine] has managed to arrive, I think we should reckon the things it has discovered—wondrous, were they from complete ignorance—as having been rightly and truly uncovered, and not the product of accident.[9]

The author of this text, whoever he was, thinks that the great degree of success medicine has achieved means that it has basically got things right. To be sure, the author is addressing his argument to a contemporary debate that has less to do with what we moderns would call realism specifically than with whether medicine as traditionally practiced is a genuine art (hence the emphasis on method in this passage). But the realist conclusion still obtains; the ideas of accuracy and "what medicine has discovered" gestured at here become most fully developed in the conclusions the author draws from his own argument a little later in the work, where he emphasizes the reality of the entities posited by the theory. He then goes on to discuss how interventions act on those entities (they become manipulable). So the body consists of a curiously idiosyncratic set of humors, namely "the salty," "the acidic," "the bitter," and so on, in one-to-one parity with the most important flavors in foods. The head draws liquid copiously up into itself because of its funnel-like structure (same with the bladder), and the lungs and breasts are sponges, sopping up fluids in the body. Obviously no modern realist would agree that this author did in fact have things right, not even close, and so we are faced with a tension between the intuitive plausibility of the miracle argument and how it appears in the Hippocratic author's hands. Let us call this *the Hippocrates problem*.

Philosophers will be immediately reminded of Larry Laudan's famous objection to the miracle argument, where he showed that in the past, some

9. *On Ancient Medicine* xii.

very wrong theories had been very successful.[10] This fact leads us directly to a conclusion that is often called *the pessimistic induction* (hereafter abbreviated PI). PI argues that if successful theories have been so often wrong in the past, then our own success can be no guarantee that we now have things right. Miracle-argument realism is then seen to be disproven historically. We need to be clear, though, that what I am pointing to in the Hippocrates problem is not just a restatement of Laudan's point. We should think of the Hippocrates problem not primarily as a single instance of wrongness (this would be Laudan's use of it) but instead as a historical instance of an invocation of the miracle argument (something Laudan nowhere discusses). The difference is subtle, to be sure, but significant. By thinking of the Hippocrates problem as an instantiation of the realist intuition, one where that intuition looks *prima facie* misleading, we do two things: we bring more clearly into consideration the historical conditions under which the argument was made, and, related to that, we shift the emphasis from worrying about its wrongness per se to thinking about what kinds of criteria we might invoke in order to block this particular claim to realism. I argue that these criteria have to meet one of the following conditions: either (a) they would have been useable by the Hippocratic author himself, or (b) we have to show that history puts us in an epistemically privileged position from which we can impose our criteria retroactively, ruling Hippocratic realism out of court even if the historical actor himself could not have seen his error.

Quantum Magnum PI?

The pessimistic induction has many minor variants in its exact wording, but all accounts are agreed on the basic argument: if you look at the history of the sciences, you find many instances of successful theories that turn out to have been completely wrong. This means that the success of our current scientific theories is no grounds for supposing that those theories are right. There are two ways of reading this. As Timothy Lyons pointed out, the vast majority of philosophers treat PI as an inductive argument (hence the name "pessimistic *induction*"). In induction, examples are collected to prove a general point, and in this case we conclude, from the fact that wrong theories have often been successful in the past, that our own successful theories may well be wrong too. Lyons, on the other hand, prefers to treat it as a

10. Laudan 1981a. For a sophisticated rethinking of Laudan's argument that targets the newer versions of realism (Kitcher 1993, and Psillos 1999) that were developed in response to Laudan's critique, see Lyons 2006.

deductive modus tollens argument, where the negation of the consequent negates the antecedent. This involves reworking PI such that it becomes a deductive argument of the form "if *p* then *q*," where a proof that "not-*q*" is then taken as disproving *p*. The argument is as simple as it is effective:[11]

> If the miracle argument is right, then all (or very nearly all) successful theories will be correct.
> But we have a long list of successful theories that are not correct.
> Therefore the miracle argument is not right.

This looks very promising, though we still need to be careful. For this is an argument that stands or falls on the solidity of its historical substructure, and yet close attention to that historical substructure may pose serious problems for the argument itself. In particular, it is not at all clear that we can talk about "success" and "failure" in blanket terms as though they were universally applicable, as though they could be divorced from the specific historical circumstances that color and inform actors' judgments of what should count as success or failure in any given experimental or rhetorical context.

Both the inductive and the deductive readings of PI emphasize the number of times in the past that successful theories have been wrong. In the wake of early criticisms, advocates of PI are now generally careful to more narrowly specify what "successful" is taken to mean (e.g., *predictive* success, to name the most common variant, or *novel* predictive success if we are being more cautious). What all variants share is that they target the intuition at the heart of the miracle argument, that success should be taken to imply some kind of match-up between theory and observation. And again, for the miracle argument's supporters, either a theory is successful because it accounts for observations or, in stricter contexts, it is successful because it generates novel predictions that are borne out in subsequent experiments or observations. We should note, though, that common versions of the miracle argument do not limit themselves to the derivation of narrow realist conclusions in laboratory settings—they are not typically about single entities or single experiments. Invocations of the miracle argument are more typically compound conclusions taken from something like "all the successes we see science and technology as having." This is a much broader

11. Lyons 2002 (with thanks to Christián Carman for drawing my attention to it). The version I give here is a paraphrase. I also realize that once it is taken noninductively, the "I" in PI becomes a misnomer, but rather than multiply acronyms (*MTP? PMT?*) I have chosen to keep the old name even when read deductively. Lyons' name for the argument, "the pessimistic meta-modus tollens," strikes me as outweighing its accuracy with its unwieldiness.

argument than just allowing J. J. Thompson a realism about electrons (and only electrons), for example, based on his own laboratory experiences with cathode rays in 1897.

We shall see in a later section how sophisticated attempts by realists to skirt some of the problems posed by PI try to look very closely at just these small-scale, local decisions about electrons and so forth in order to try to determine what the minimal commitments to theoretical posits might be in the light of experimental successes. At the same time, though, if we pay careful attention to what criteria for success are used by historical actors on the ground, we find that it becomes increasingly difficult to generalize about what the successes of phlogiston, caloric, the luminiferous ether, and arterial pneuma all have in common. They may all be seen as wrong now, but like Tolstoy's unhappy families, they are all wrong in their own ways. The observational phenomena, the theoretical considerations, the anomalies, the hands-on laboratory challenges, the larger conceptual frameworks—all of these are very different for phlogiston than they are for Galenic pneuma. Indeed, as we pay closer and closer attention to historical instances of inference from successes to conclusions about entities or structures in the world, we find it more and more difficult to say why observation and success *in general* cannot be trusted in the sciences, and find ourselves instead limited to saying a lot of little things about material assumptions in optics, experimental results in calcination, or success criteria in a particular physiology laboratory. This raises a much more general (and significant) point: if we look closely at where the Hippocrates problem went wrong, we find that the specifics of the "success" that the author was pointing to do not—indeed cannot—map one-to-one onto the specifics of other realism-provoking successes, then or since. Indeed, what counts as success in one science or in one time and place all too often does not even line up in kind with that of another. It begins to look as if there is no such thing as simple "predictive success," but instead many individual successes. This is an important point that has often been underappreciated: the predictions that stand out to convince an eighteenth-century chemist of the existence of phlogiston are very local. They have to do with specific challenges faced by the experimenter and the theorist under specific conditions. Moreover, not all successes are equal. Depending on the challenges and the results, individual successes are ranked by participants, some more impressive than others. How successes are ranked, which ones are really convincing, all this is again very local. Finally, the kinds of problems that unobservable entities run into when they come to be questioned are similarly individual. If we look closely at the eventual experimental disproof of the luminiferous ether, and how and why it convinced a community of scientists at the time, we find that the complex of its convincingness is not at all the same as that for the caloric theory of heat, or, for that matter, for the four humors. The ways to kill a theory depend very much on who is to be convinced and why. Historical conditions matter, and historical conditions are local.

Can We Avoid the Problems History Poses?

Over the years, PI and other objections have forced philosophers to refine their arguments for realism, and two broad (and often overlapping) strategies have evolved in order to explain what historical actors were doing that led them to the wrong conclusions about the world, and how we can (or at least hope to) avoid making the same mistakes. The first strategy in effect essentializes the modern and historical sciences, and says that there is some quality the modern ones have that the historical sciences were missing, and this quality gives our science a privileged window on the world that theirs was lacking. The second strategy focuses instead on the ways in which people draw conclusions from the success of their theories, and says that the problem lies in which parts of the theories they take to be reality-indicative. Historical actors went wrong because they assumed that success proved everything about their theories, whereas in fact success can only prove one small aspect of larger theories (not everyone agrees on which aspect). I will argue that neither strategy is successful.

First Strategy: We Have Something They Didn't

In the face of antirealist objections, Philip Kitcher complained that "there are almost as many versions of realism as there are antirealists, each ready to supply a preferred characterization before undertaking demolition."[12] I would counter the implication here that antirealists are the multipliers by pointing out that the proliferation of realisms is not entirely their doing: realists love a subtle distinction as much as anybody in the business. Just parsing the many published versions of realism that apply to the questions we have been asking of Roman science in this book, and leaving out the various ethical, political, and aesthetic versions of realism, we get this (incomplete) list:[13]

12. Kitcher 2001, 151.

13. Carman 2005 actually argues that, combinatorically, there are at least 1,111 possible kinds of scientific realism.

semantic realism
metaphysical realism
scientific realism
eclectic realism
critical realism
critical scientific realism
deployment realism
eliminative realism
entity realism
structural realism
internal realism
epistemic realism
possible-worlds realism
naturalistic realism
objective naturalistic realism
modal realism
extreme modal realism
axiological realism
methodological realism
naïve realism
Platonic realism
constructive realism
semi-realism
and even *real* realism

One handy subdivision of key types that distills this proliferation down to something more suited to our purposes comes from Stathis Psillos. As he summarizes the main families of positions, we get:

(1) Metaphysical realism: the world has a definite and mind-independent structure.
(2) Semantic realism: scientific theories should be taken at face value.
(3) Epistemic realism: mature and predictively successful scientific theories are (approximately) true of the world. So, the entities posited by them, or, at any rate, entities very similar to those posited, inhabit the world.[14]

14. The summary comes from Psillos 2005. The original argument is in Psillos 1999.

It is in Psillos' (3) that we see the most often-invoked keywords for neutralizing the standard antirealist uses of history, and the best contenders for cordoning off the Hippocrates' problem. These are the magic words *mature,* and *predictively successful,* both of which are taken to apply to the modern sciences (or at least the legitimately realism-generating modern sciences) in ways that they do not or cannot apply to the premodern.

But close attention to the Hippocrates problem should give us pause here. One point that emerges is that we today are not alone in being on the crest of the wave, so to speak. All historical actors live at the ends of long lines of development, and it seems that thinking that one has come an impressively long way may therefore be anyone's prerogative. This means that in order to avoid the unidirectionality-of-vision problem (the underappreciated but nontrivial objection that although we have clearly come a long way, we have no idea how wrong we may look to have been in a hundred or a thousand years), we need to find metacriteria for sufficiency of both predictive success and maturity that can be dehistoricized—and I will argue that this is not possible.[15] Lacking metacriteria that define how an idividual actor at a specific point in history could know whether his or her criteria for realism-triggering predictive success were sufficient, the historical actor's criteria for realism-triggering sufficiency—and indeed any criteria for bracketing false historical realisms—fall back to a version of the *ad novitatem* fallacy, the logical fallacy of grounding an argument in the claim that newer is just better than older. The question we need to raise, then, is whether there are any metacriteria for deciding whether one's own science is *sufficiently* mature and predictively successful that do not fall prey to an *ad novitatem* fallacy. I submit that there are not.

First, to deal with predictive success. There are, we should say at the outset, many worries about prediction as a criterion for the acceptability of a theory. We should take seriously the many studies that have shown that in practice it is in fact explanatory breadth that does much of the work that often gets falsely credited to prediction.[16] Like the proliferation of

15. Stanford 2006 uses historical examples to unpack a similar set of issues as what he calls the "problem of unconceived alternatives," which he directs at both PI and what philosophers of science call "the underdetermination of theory by evidence." His emphasis, however, steers away from traditional readings of PI and underdetermination to focus instead on how practicing theorists conceive of (and fail to conceive of) explanations. My own argument more directly targets realist attempts to corral PI by defining some essential truth-preserving feature that modern science has and ancient science lacked. Compare esp. Stanford 2006, 45f.

16. Brush 1989, 1994; Laudan 1981a; Howson and Franklin 1991; Collins 1994; Achinstein 1994.

observations-nobody-has-had in connection with garlic-magnet antipathy, the actual epistemic roles of predictive success tend to be overemphasized by actors in the field. Nevertheless, philosophers continue to struggle to find some way of salvaging a ground for the admittedly common intuition that a novel prediction of the "I told you so" variety should carry more epistemic weight than the (potentially ad hoc) incorporation of known or new data into a theory.[17] And the search for a justification of that intuition continues apace. Whatever the difficulties may be here, let us grant the significant psychological fact that the success of a prediction is widely thought to be convincing. We should also note that this has been the case for a very long time.

Galen, for example, made spectacular use of the psychological force of successful prediction in outmaneuvering his competitors in the great game of patronage at Rome.[18] In his treatment of the deathly ill peripatetic philosopher Eudemus, Galen tells us how he proved to everyone the correctness of his own theory of medicine over its competitors:

> ἐκ γὰρ τῶν τριῶν τεταρταίων τὸν μὲν πρῶτον ἀρξάμενον ἐν τῇδε παύσασθαι τῇ ἡμέρᾳ προειπὼν ἐθαυμάσθην· ἐπεὶ δὲ καὶ τοῦ δευτέρου τὴν προθεσμίαν τῆς λύσεως ἠλήθευσα ἅπαντες μὲν ἐξεπλάγησαν· ἐπὶ τοῦ τρίτου γοῦν ἀποτυχεῖν με τοῖς θεοῖς ηὔχοντο. παυσαμένου δὲ κἀκείνου κατὰ τὴν ὑπ᾽ ἐμοῦ προρρηθεῖσαν ἡμέραν οὐκ ἐπὶ ταῖς προρρήσεσι μόνον ἀλλὰ καὶ τῇ θεραπείᾳ δόξαν ἔσχον οὐ σμικράν.

> Because I had predicted the exact day on which the very first of the three quartan fevers would end, I was looked on with wonder. But when I was also right about the day of the loosening of the second, everyone was dumbfounded. Regarding the third, [my opponents] prayed to the gods I would fail. But when that one also ended on the predicted day, I got no small reputation—not only for my foresight, but also for my treatments.[19]

This is no small feat.[20] The combination of three distinct but overlapping cyclic fevers looked chaotic and incomprehensible to Galen's competitors.

17. See, e.g., Musgrave 1974; Maher 1988, 1990; Mayo 1996; White 2003; Worrall 2006.

18. See Mattern 2008 esp. chap. 3. On patronage generally, see Saller 1982; Wallace-Hadrill 1989; Canali de Rossi 2001; Verboven 2002.

19. Galen, *Prog.* 2.27.5

20. I recognize that we need to treat any claim by Galen on the subject of his own reputation very carefully. In this instance, though, it is enough that he hoped his reader to be convinced by his predictive successes. We might add that as tedious as his self-serving prose can sometimes seem, he was in fact spectacularly successful in his career—his ability to wow audiences was not *all* in his head.

Throughout the initial period of his consultation with Eudemus, months before the fevers broke as he had so "long ago" foretold,[21] Galen made successful prediction after successful prediction about relapses—even down to the exact hour—and he was the only physician to correctly predict how Eudemus would respond to the various treatments that his other doctors were trying in desperation. When the fevers finally made their last stands, Galen was able to predict not only that these *were* the last stands, but also precisely how and when the morbidity would be finally evacuated from Eudemus' body. At every stage of the progress of these diseases, Galen had complete and highly detailed predictive control—and here it does not matter whether we agree that Galen was in fact predictively successful, but only whether Galen and his contemporaries believed him to be. Notice too that the compounding predictive successes lead not only to the conclusion that Galen is good at prediction (which would be inductively sound enough), but also to the conclusion that Galen is good at treatment, which does not strictly follow.[22] The intermediary proposition can, however, be supplied: Galen's contemporaries are supposed to conclude that he is good at treatment because he understands the body, and his understanding of the body is shown by his predictive success. Indeed, Galen himself fills in the details of his prognosis over the next few paragraphs and shows that it is his knowledge of pulses in particular, and of how specific internal bodily processes associated with disease affect pulses, that allows him to make his accurate predictions. Certainly, the complexity, difficulty, and richness of Galen's theory of the pulse is impressive (I count at least six dimensions in which pulse is measured, for example).[23] That he himself thinks his success in this case is anything but trivial is shown by the degree of praise he claims his contemporaries lavished on him and the scorn with which they treated his opponents. As with the bare version of the miracle argument, we would need to specify dehistoricized criteria for the question of "how much predictive success is enough?" if we are to deny Galen a legitimate realism based on his impressive results. Galen thought he had sufficient success; we think we do. And it appears that the line is being drawn only by an equation of "now-ness" and "rightness" in both cases.

21. *Prog.* 3.18.1

22. Galen does not detail what treatments he may actually have performed on Eudemus (although he hints that he visited Eudemus frequently and on one occasion gave him theriac). The main therapeutic method he discusses explicitly seems to have been simply to steer Eudemus away from the other doctors. Importantly, though, Galen's primary tool in convincing Eudemus to distrust the other doctors was his own predictive success.

23. On Galenic pulse theory, see Kuriyama 2002; Boylan 2007.

To see why this is so, we need only think of some modern predictions that we might invoke as paradigmatically convincing to the modern successful-prediction realist, and that might be frameable in such a way as to establish a fundamental difference between our predictions and Galen's. Perhaps we could point to the prediction that led to the discovery of the planet Neptune, or the prediction we hope will be fulfilled with the discovery of the Higgs boson. Imagine, then, delimiting some realism-triggering criteria for prediction that would mark a definitive line between Galen's and these successes. What might it look like? We could work out some careful rules around lawlikeness. We could carefully delimit experimental conditions to keep Galen's trussed-up pigs out and the Large Hadron Collider in. Perhaps we should appeal to the success of our manipulation of unobservable entities, which success we must then distinguish from Galen's belief in his own successful manipulations of pneuma. Some carefully formulated mathematization may enter the picture, one cleverly designed to keep Ptolemy at bay as well. Let us suppose we could, at least tentatively, carve out such a criterion. What would it look like to Galen or Ptolemy?

It would, I submit, have been impossible for them to conceive of. Not because they had some fundamental way of thinking about the world that was different from ours, but simply because of where they existed in time. A Roman could not have come up with a list of criteria that predictions would need to meet in order for realism to be acceptable that would have let CERN in and left the Romans' own science out. To do so he would have to have invoked incomprehensible rules, unimagined entities and structures, impossible processes (we might call this the problem of unconceived methodological alternatives). This is because the rules themselves that we would want to invoke depend on the outcomes of a particular historical trajectory. We may well think it a good and impressive trajectory, but that does not get us any farther than the Hippocrates problem. This is because the ability to make the judgment we would need, indeed the ability to even conceive of the conditions of the judgment, depends on the contingency of inhabiting a specific historical time frame—which very conveniently just happens to be our own. But how can we know that the historical time frame we do in fact inhabit is, at long last, epistemically adequate? What can we employ beyond simply that it all looks impressive to us now? Part of this worry is that for our confidence to be stronger than Galen's or the Hippocratic author's had been, we need some way of underpinning our confidence that we will not be faced with a similar set of historical changes sixty or six hundred years hence, changes that will render our own realism premature just as happened to the author of *On Ancient Medicine.* Again, this underpinning of our confidence needs to distinguish itself from the bare it-looks-sufficiently-impressive claim; otherwise we are doing nothing more than rehashing an old (and, the realist would have to say, clearly fallible) argument that our criteria are sufficient because, well, they are better than we used to have.

But the modern realist has one more card up his sleeve: *maturity.* He can still try to claim that only mature scientific theories are (approximately) true of the world, so as to limit the genuine acceptability of historically posited entities or structures and thus avoid a realism of psychic pneuma, garlic-magnet antipathy, cosmic harmony, and all of the other discredited characters filling the pages of this book. In order to make this move, though, he needs to block the road to the *ad novitatem* problem that has been dogging predictive success, for if it turns out that all he can mean by "maturity" is "what we have now," then we are once again in the mire.

So what does maturity mean? Everyone who invokes it agrees that modern physics is mature, and in a great many instances that is exactly as far as the analysis goes: mature sciences are like contemporary physics in some vague and unspecified way.[24] If we cast around for books and articles with "mature science" in their titles, we discover a rather common desire to establish whether this or that newer discipline or subdiscipline is mature:[25]

"Visions of a Mature Cognitive Science?"
"Coming of Age in Quantum Optics"
"Towards a Mature Social Science"
"Is Medical Informatics a Mature Science?"
"Is Population Ecology a Mature Science?"
"Meteor Astronomy: A Mature Science?"

and from an chippy entomologist in 1960:

"Mature Science—Retarded Profession."

If we mine these articles for conceptions of what maturity might mean, we get a range of options, where one of the following is variously taken to suffice: sufficient longevity (importantly: measured in decades rather than centuries); clear disciplinary boundaries; predictive ability; or explicit attention to issues of measurement. In one instance the occurrence of a "Kuhnian

24. Putnam 1978, 21; Musgrave 1992, e.g.

25. The examples come from: Sanocki 1996; Stenholm 1997; Olson 1983; Friedman and Abbas 2003; Getz 2003; Beech 1988; Rogers 1960.

revolution" is taken as singly sufficient to make a science (meteor astronomy) mature.[26]

Among philosophers, the situation is, alas, not much better. Invocations of maturity are legion, but the place-marker held by the adjective is almost never spelled out. "Mature" is often used simply to mean "legitimate," roughly implying "recent," though the latter cannot be taken as definitional without once again hitting either multiple historical realisms or the *ad novitatem* problems we saw above. Indeed, as often as not in the philosophical literature, the adjective *mature* can be deleted with no perceptible change in meaning at all. It is in this sense a mere caveat among a discourse community, used to mark the fact that the speaker knows history can be a problem but wants some things to be immune to it—maturity then becomes little more than a kind of talisman.

Where we do get hints at real philosophical criteria in play, there is often not enough to separate our science from Galen's or Ptolemy's. So Kuhn equates a mature science with a science that has a paradigm, but then he doesn't rest very much on the idea (and in any case, Kuhn explicitly gives Ptolemy maturity).[27] Putnam is fond of the word, borrowing it from Richard Boyd, but again it basically means nothing more than "legitimate."[28] John Worrall simply equates mature science with predictively successful science,[29] and Barbara Tuchańska says that a mature science is "one in which empirical occurrences are results of theoretically permeated experimental operations,"[30] none of which eliminates enough to avoid the problems we have already seen. Ptolemy and Galen, for example, are still very much in the game. Perhaps the most detailed discussion, and one we commonly run into when we chase down the footnotes to maturity, is by Wadysaw Krajewski. On a charitable reading, Krajewski draws the line relative to types of revolution: sciences that are prone to having their basic entities or theories dispersed and replaced in a revolution are immature, and sciences whose revolutions are such that they are able to incorporate prior theories as approximations or limiting cases of the new, are mature. It is hard not to read this as simply meaning that sciences that are prone to being wrong are immature, and those that are approximately right are mature. Once again, "mature" seems simply to equal "not being subject to historical objections."

26. Beech 1988.
27. Kuhn 1970, 21–22.
28. Putnam 1978, 20–21.
29. Worrall 1989, 113f.
30. Tuchańska 1992.

Moreover, there are no criteria on offer for judging whether a given change is really of the first or second sort (how much incorporation is enough? of what kinds of entities? laws? equations? . . .). Perhaps more importantly, we are also not told how we might know whether our own scientific disciplines are insulated from the major type of change in the future. When it comes to offering concrete examples of maturity, Krajewski simply appeals to the following unhelpful criterion (and we should seriously worry about the fact that he offers this criterion definitionally): "An immature science grasps wrongly the essence of its objects, a mature science does it correctly in rough outline."[31] So the Copernican revolution marked the transition to maturity for astronomy because whereas "the [geocentric cosmology] does not grasp the structure of our planet system properly, the [heliocentric cosmology] does so in a rough outline." Likewise, "Aristotle did not grasp the essence of movement properly, Galileo did."[32] But again, all anyone has to work with from *within* a historical time period is an impression that nothing major has happened for a while, which seems a terribly dubious ground for realism. Worse, if we try to put a time frame on the stability criterion (how much time needs to have passed since the last major upheaval in order for us to call a science "mature"?) then we are faced with the fact that all our mature sciences are dwarfed in their longevity by virtually any ancient one. So again, Galen was right to be a realist about the four humors, septal pores, and psychic pneuma.

Where does this leave us? Look back at Psillos' definition of epistemic realism: "Mature and predictively successful scientific theories are (approximately) true of the world." Neither maturity nor predictive success have useable criteria on offer for eliminating Galenic or Hippocratic realism about entities that our theories now reject. Nor do the prospects look promising for either,[33] simply because we can find historical examples that undermine the idea that the reason our science is right is because it is impressive to us. No doubt it is impressive, but smart ancients seem to be just as impressed with their sciences, and for the same reasons. And being impressed is not the same as being right.

31. Krajewski 1977, 89. Laudan 1981a, 34, points out that even Krajewski admitted this pairing of criteria to be tautological.

32. Krajewski 1977, 89. His claim on page 90 that Aristotle and Ptolemy were also immature scientists because they did not critically examine the theories they superseded is so bizarre as to merit no rebuttal.

33. For more pessimism about maturity, see Achinstein 2002.

Second Strategy: The Curate's Egg

There is another set of broad strategies for dealing with pessimistic induction that, like maturity and predictive success, carries historiographic implications. Timothy Lyons has called this family of strategies *deployment realism,* but I prefer to think of it as *curate's-egg realism,* recalling the 1895 Punch cartoon (see fig. 9.2) in which the reaction of a nervous young curate to being served a bad egg by the bishop is to stammer that "parts of it are excellent!"[34]

The idea behind these versions of realism is to try to get around the problems posed by Laudan's wrong-theories-are-successful argument by emphasizing that theories are rather big and sometimes nonhomogenous things, and that we need not be realists about the whole package.[35] Instead, we can be realists about just those parts of the theories that genuinely contributed to success, leaving behind all the extra theoretical baggage that only seemed to be contributing, but wasn't. We keep the genuinely right bits, and throw out the wrong ones. We can simultaneously see what went wrong with historical cases of mistaken realist inferences from the successes of wrong theories: those actors chose the wrong *parts* of their theories to be realists about. We can then treat all of Laudan's wrong theories as only "mostly dead," and isolate some parts of them, those that really were responsible for the successes, as basically right. The trick, though, lies in picking out the right parts. Moreover, if we try to open the historical window wider and frame the Roman sciences within curate's-egg narratives, we run into considerable difficulty. On the one hand, it is exceedingly difficult to say what little realist gems we can tease out of the Galenic theory of the humors or out of Ptolemaic cosmic harmonics. And on the other hand, as we've seen again and again in this study, historical actors (ourselves included) frequently misconstrue even the most basic categories in observation and theory.

As was the case with predictive success, we should need some way for historically situated actors to themselves know which bits are genuinely contributing to their successes, and which bits are merely along for the ride. Without a means of doing so, we can only be realists retroactively, so that we—now—can be realists about the parts of the luminiferous ether theory

34. *Punch,* November 9, 1895. Cartoon by George du Maurier. I thank Jay Foster for drawing my attention to the example.

35. For the argument in favor, see Worrall 1989; Kitcher 1993; Psillos 1999. See also Krüger 1981. For a critique of Psillos and Kitcher (on which my own partly builds) see Lyons 2006.

Fig. 9.2. Bishop: "I'm afraid you've got a bad egg, Mr Jones."
Curate: "Oh, no, my Lord. I assure you that parts of it are excellent!"
"True Humility," by George DuMaurier, in *Punch*, November 9, 1895.

that were the right ones, even if the people deriving the successes from the theory at the time had no way of knowing which were the essential parts. There are various strategies for dealing with this problem. The most common approaches try to get around it by arguing that legitimate realism-triggering hypotheses can be identified as only those playing some particular role in the derivation of a theory's predictive success. One version of this approach is taken by Psillos, who argues that we can be realists about those parts of a theory that were *indispensable* to its predictive successes, where indispensability includes (in addition to its obvious sense) the caveat that there be "no other available hypothesis" that could substitute (since substitutability is dispensability).[36] In light of the argument I developed above, though, the Hippocrates problem means that "availability" by itself is too narrow, too perspective-based, too historically contingent (available when? available to whom?), to eliminate promiscuous realisms. To see this, think of how differently Hippocrates would evaluate which of his own hypotheses

36. As Psillos (1999, 110) puts it: "When does a theoretical constituent *H* indispensably contribute to the generation of, say, a successful prediction? Suppose that *H* together with another set of hypotheses *H′* (and some auxiliaries *A*) entail a prediction *P*. *H* indispensably contributes to the generation of *P* if *H′* and *A* alone cannot yield *P* and no other available hypothesis H^* which is consistent with *H′* and *A* can replace *H* without loss in the relevant derivation of *P*." For a critique, see Lyons 2006.

were indispensable (or to use Kitcher's criterion: *nonidle*), versus how we might evaluate and explain his success. Consider as an analogy how Kitcher handles Fresnel's early nineteenth-century prediction of the so-called Poisson bright spot in optics. Fresnel's prediction relied on a luminiferous ether, a physical medium that functioned like a hyperthin version of air or water for the propagation of light waves, but a medium no scientist today believes to exist. Fresnel's mathematics is still used in some modern optical contexts, however, so that aspect was clearly not idle, and the math continues to generate the right dispersion pattern even without the ether in which Fresnel initially embedded it. Kitcher may be right that this addresses Laudan's worry about how wrong theories can get things right, but it cannot eliminate the Hippocrates problem, for we are not given criteria by which Fresnel himself could have seen the superfluity of the ether. Indeed, the requirement for the isolation of Fresnel's ether hypothesis as idle, was the development of post-Fresnelian optics. We can, in short, be curate's-egg realists only in retrospect: we now can see which parts were right in previous theories. Moreover, as Kyle Stanford has argued, the very standards by which we judge parts of past theories to have been right or wrong, and by which we judge individual commitments to have been nonidle or indispensable, are *entirely* the products of which aspects of those theories survived historically, which criterion could never have been employed by the historical actors themselves.[37] Some realists may be happy with such a retrospective realism, but the problem then arises that we ourselves cannot be realists about our own science; only future scientists can be realists in retrospect about the parts of our science that turn out to have been nonidle. Not only this, but we get an infinite regress: if only future scientists can determine which of *our* hypotheses are nonidle, then only future scientists are in a position to truly determine which of *Fresnel's* had genuinely been nonidle—but then again, only future-future scientists can determine which of the future scientists' hypotheses were nonidle, ad infinitum. Without either an end to history or the assumption of a gods'-eye view, we have no grounds for curate's-egg realism from within history itself.

Other Ways Out

One final proposal for determining which parts of a theory can be expected to pass the test of time is Anjan Chakravartty's turn to detection properties, which are those properties by means of which an entity is detected, as

37. Stanford 2006.

the click on a Geiger counter allows us to detect beta particles.[38] We can, Chakravartty says, be realists about the minimal concatenation of causal properties and structures that must underlie detection events (we might loosely say "observations"). Given a theory (for Chakravartty, usually expressed as an equation) that accounts for some detection event, Chakravartty then asks what the minimal interpretation is that we have to give to the terms of the equation in order to make realist sense of it with respect to the detection event itself. But two worries present themselves. The first is that there is a perennial problem with the determination of relevant detection events and their relevant aspects. How do we isolate signal from noise? Coincident phenomena from coherent? Surely there are strategies for doing both, but there is an epistemological asymmetry that limits the confidence we can have in such strategies depending on whether they yield positive or negative results. We can perform any number of careful tests to try to ensure that the detection properties we are pointing to as realism-triggering are robust, genuine, and discrete phenomena, but in the end they have the same epistemological status as Popper's not-yet-falsifieds. That is to say, we have some degree of confidence in them, but at the same time we should not be surprised to find that, perhaps many years later, the very nature of what we thought we or our machines were seeing has to be reevaluated from the ground up. If we look back to Galen, say, it becomes abundantly clear that what he would have listed as relevant and robust realism-triggering detection properties were both deeply embedded and ultimately nonlasting. He really thought he *saw* humors being evacuated. Even if we could narrow down his observation claim to just the colors and textures that were before his eyes, urging him to resist the identification of the group of phenomena as a humor, we are still faced with the fact that he is not, by modern standards, even looking in the right place, in the sense that the relationship between the evacuation and the cause of the disease is, we should say, fundamentally misascribed. Perhaps even more worrying, the careful examination of causal properties behind the structures of the body that we see in Galen's *De usu partium* very prominently includes one type of cause, the final cause, that modern science may not even easily allow a translation of. Finally, there is the problem of how we deal with the messiness that inheres in other people's reports of observations (which implies other people's prefiltering of data) in testimony. No one is suggesting we need to only be realists about detections we ourselves have performed, but then I cannot see how to prevent some new version of garlic-magnet antipathy from get-

38. Chakravartty 1998, 2007.

ting a foothold. To be sure, Chakravartty is happy enough with the prospect (or even the certainty) of much more radical theory change than most realists are. But I wonder how much we have to modify his main argument if we break the Fresgiston Pact—as I think we must—and bring to bear some truly weird science.

The second worry about Chakravartty's position is the slipperiness of the phrase "minimal interpretation." The problem is not with the idea motivating its use, so much as with how historical actors would have interpreted it. We have already seen how the historical record highlights that scientists regularly commit to considerably more of their theories than later history has shown to have been strictly justified (we now can see that they were not, in practice, really minimalists). But as Stanford has shown, these very scientists regularly go so far as to explicitly defend those same larger commitments as *necessary* to the laboratory outcomes (predictions, etc.) that they are witnessing.[39] To say that this problem would disappear if our historical scientists would only have committed to a minimal interpretation of their theories is no different than Kitcher's insistence that the problem of PI would be solved had past scientists only committed to the nonidle parts of their theories. And in exactly the same way, our historical actors seem to regularly misconstrue what we later take to be the minimum interpretation to which they should have committed. That is to say, minimalism is merely another curate's-egg attempt to cordon off the history of the sciences, and it faces the same objections that indispensability and nonidleness did above.

Turning back to Psillos, the solution I propose lies in adding an explicit qualifier to his (3) to produce:

(3a) *People are justified in believing that* scientific theories are (approximately) true of the world. So, *they are justified in believing that* the entities posited by them, or, at any rate, entities very similar to those posited, inhabit the world.

This is a perfectly defensible position, but it does limit what we can know about the world in ways that will not satisfy all realists.[40] In particular, it

39. Stanford 2006, chaps. 6–7.

40. Psillos, for example, who gave us our threefold division in the first place, may occasionally look to be talking in terms of justification, but it is clear that he wants a stronger account of truth than justifiability alone can afford. So he rejects the set of qualifications in (3a) on the grounds of incompatibility with the "are (approximately) true" part of the formulation. See, e.g., Psillos 1999. We can see the full import of this in his theory of reference, where "luminiferous ether" refers (because there is a *real* entity, the electromagnetic field, to which it refers)

succeeds in getting us out of the dilemma we have been facing, but at a cost: we have to allow our historical actors their realisms too. Moreover, we have backed away from a direct account of realism as an ontology to an account of the justification for realism. Rather than proving an ontology itself, we seem to be limited to the argument that scientists are justified in positing that ontology. In the light of the analysis of this book, I claim that this is not too high a cost to pay, and in any case it may be the only way of salvaging realism in light of history.

In the next chapter, I will add to this argument by showing why some degree of realism is inescapable for the historian, even if (as emerged in the wake of Kuhn's work) there are some real historiographic advantages to relativism. I will then try to sketch a version of realism that is consistent with the historiographic insights at the heart of the post-Kuhnian move.

---

but "phlogiston" does not (Psillos 1999, chap. 12), which position would be untenable without imputing considerably more to the epistemic realist's entities than mere justifiability of belief in them.

CHAPTER TEN

Worlds Given, Worlds Made

In this chapter I argue that if we cannot deny our historical actors their realism, then what it means for something to be "real" would seem to come under direct threat. Relativism (using that term very loosely) or some other version of antirealism may be tempting as an easy way out here, but this move would simultaneously eliminate one of the historian's most fundamental tools for coming to grips with past understandings of nature. I instead sketch another route, one that takes as its base a weak realism coupled with a coherence-based epistemology. This approach has two virtues. On the one hand, it makes consistent philosophical sense of the cognitive dissonances that underlie and motivate many of the topics covered in this book. It also offers a philosophically consistent ground for historiography, one that is in keeping with the important insights at the heart of the post-Kuhnian move while avoiding what has often seemed an inevitable attendant relativism. There is, of course, a price to pay. By making this move we put significant pressure on what we might mean by "real," pressure that can only be relieved by looking very closely at the juncture between "real" and "true."

What's in a World?

The fact that natural phenomena are understood very differently over time is the most basic starting point of any investigation of the history of the sciences. But it poses an age-old historiographic problem. If the historian understands the phenomena of nature as being real, out-there-in-the-world things, structures, processes, and so on, then it seems to be the case that any difference of opinion about what constitutes those phenomena can only happen if at least one of the differing parties is fundamentally *wrong* about what is happening. Certainly this was the approach that characterized the earliest work in the history of ancient science, and it had as a fellow traveler the corollary that the historical trajectory of the sciences was one of convergence on the truth: we correct errors over time, and are therefore coming ever closer to what nature really is ("convergent realism"). But, particularly in the wake of Thomas Kuhn's groundbreaking *Structure of Scientific Revolutions,* this approach began to appear misleading, at least partly because it inherently misrepresents much of what historical scientists saw themselves as doing. The post-Kuhnian move emphasized seeing past scientific practice as embedded in historical social and intellectual cultures, and emphasizing a less anachronistic approach to problem sets and what we might call intellectual tool kits. This had a number of important advantages, and avoided the most obvious positivist pitfalls of the older approach, but it did not do away with all of the old demons. The turn away from convergent realism allowed the fields that studied the historical sciences to develop some very powerful alternative approaches to the historical material, and at the same time provided a more accurate and dynamic picture of the actual practices of the historical sciences, where scientists were no longer treated as though they were hermetically sealed off from their wider cultures, places, and priorities. One consequence of this approach was to emphasize the ways in which entities and processes are *constructed,* to use one kind of jargon,[1] and at the extreme end, the claim was that these entities and processes were *only* constructed. In Nelson Goodman's provocative framing: *worlds are made.*[2]

The question of whether worlds are made or whether the world is something out there to be discovered by correct application of reason and experiment is properly a philosophical one, and is accordingly not often given extended consideration by historians of the sciences. One important point I want to make in this chapter is that the bracketing of this question by historians is belied by an important part of their historiographic method. The problem has to do with the ways in which historians' experiences and understanding of the world they live in—its entities, structures, processes, possibilities—figure into their project of understanding ancient experiences of that same world. Looking out to the world they know in order to then try to understand the world that the Romans knew is to make some not inconsequential assumptions about what the world is and how it stands in relation to its observers. I argue that only by taking a (weak) "realist stance" can we be consistent with historians' actual use of their sources—most sig-

1. A good survey (and cogent critique) can be found in Hacking 1999.

2. Goodman 1978.

nificantly this "third source," their experiences of the world—while still recognizing the important historiographic insight at the heart of the post-Kuhnian move.

Kuhn's World

Over the years, one particularly busy line of work in the history and in the philosophy of the sciences has been centered on untangling what happens when theories, or whole groups of theories, are rejected in favor of new ones. In its most common formulation, this is the question of "scientific revolutions" that got its greatest momentum in the wake of Kuhn's work.[3] One of the upshots of Kuhn's ideas about paradigms and the incommensurability that attended theory change was his talk of scientists "living in different worlds" before and after a revolution. This was his provocative original formulation:

> Examining the record of past research from the vantage of contemporary historiography, the historian of science may be tempted to exclaim that when paradigms change, the world itself changes with them. Led by a new paradigm, scientists adopt new instruments and look in new places. Even more important, during revolutions scientists see new and different things when looking with familiar instruments in places they have looked before. It is rather as if the professional community had been suddenly transported to another planet where familiar objects are seen in a different light and are joined by unfamiliar ones as well. Of course, nothing of quite that sort does occur: there is no geographical transplantation; outside the laboratory everyday affairs usually continue as before. Nevertheless, paradigm changes do cause scientists to see the world of their research-engagement differently. In so far as their only recourse to that world is through what they see and do, we may want to say that after a revolution scientists are responding to a different world.[4]

I take the insight at the heart of this way of seeing scientific change very seriously, and it has been a guiding principle of this book so far. Nevertheless, Kuhn's talk of worlds, if he means it literally, would have major and undesirable consequences. Does he really mean to imply that because they have different theories, different tools, and different questions, the Romans

3. Kuhn 1962.
4. Kuhn 1970, 111.

really live in a different world than we do? If we look at the care with which he picks his language in this passage, it would perhaps seem not: we "may be tempted," he says, "it is rather as if," "we may want to say"—and of course there is the crucial use of *seeing* and *responding to* a different world rather than "living in it."[5] On the other hand, it has struck many commentators that Kuhn's idea of incommensurability actually commits him to a stronger relativism than he is allowing himself here, and that the attempts to water it down with qualifications in this passage are inconsistent with his larger position. For Kuhn, scientific revolutions are such deep conceptual ruptures that the people on one side think and talk about the world in ways that are incommensurable with, and possibly not even fully translatable into, the terms and concepts of those on the other. As he did with the word "paradigm," which had previously been a term most commonly found in old Greek and Latin grammars, Kuhn is here harking back to his classical education and borrowing the word "incommensurable" from Greek geometry, where it refers to quantities that cannot even be expressed in a given number system. Pi, for example, cannot be expressed in any number system built on our number 1 as the unit (this is why the decimal runs on forever). Pi is incommensurable, which is Latin for "not co-measurable," with the number 1. Depending on how we read the incommensurability between pre- and postrevolution sciences, we may be forced into relativism and a strong reading of what it means to "live in a different world."

Finally, we may find the seeds that undo Kuhn's moderation in the phrase "in so far as their only recourse to that world is through what they see and do." Indeed, by the end of the chapter of which the paragraph above is the opening salvo, and to an even greater extent in Kuhn's later work, we see many of the qualifications being routinely dropped, with world-talk being embraced more openly and more literally.

Ian Hacking proposed a way out of the resulting dilemma: there are two definitions of "world" that we need to steer between here.[6] One is the "world of individuals": individual horses, individual atoms, individual patients. An individual horse standing out in a field does not go away when Kuhn's world changes. It is the *other* sense of world that is affected, the "world of kinds." So we once thought garlic was one kind of thing, sympathetic, and now we do not. Garlic has become a different kind, and that has important consequences for our experiences of it, for its uses, its possibilities, its very place in our world. So also the death of sympathy as a kind-

5. This section is indebted to Hacking 1993.
6. Hacking 1993.

term has profound effects for how we experience the interrelations between things—and not just sympathetic things like magnets and goat's blood, but all things, which have now lost one primary possibility of interaction. The effect is profound enough to be classed as world-changing in Kuhn's sense. But Hacking wants to say that Kuhn's sense of worlds only makes sense if we read it in terms of kinds, not of individuals.

This sounds like a fruitful way of reading Kuhn, and certainly it promises to clarify Kuhn's hesitancy in *Structure* to commit to a literalist version of living in a different world. But Kuhn himself, in a 1993 commentary on Hacking's paper, objects.[7] Hacking's kinds, he argues, don't do enough work. Instead Kuhn chooses to talk of concepts and conceptual change between worlds. He still often uses the word "kind," to be sure, but he nuances it differently than Hacking does. Kuhn points to the ways in which concepts come bundled together: we don't generally learn just one at a time, but instead get "reptile" or at any event "nonmammal" when we learn "mammal," or else we get nests of theories and relationships in which the concept is embedded, as "force" comes with "mass" and with examples of relations between these. This bundling has deep effects on shaping our expectations about how the world works, and because of this, on the meanings of the kind-terms we employ. But if we mean something different by a kind-term than someone else does when they use it, not just in the reference but in the manifold implications that the reference carries with it (as is the case when I say "magnet" versus when Plutarch does), are we not talking about different things that inhabit different worlds? My boundaries for the object—even for the individual magnet I am actually pointing at—are here; his are there. The thing I am talking about has all of these relations; the thing Plutarch is talking about has those. Now mustn't we say we are living in different worlds in a stronger sense than Hacking allows?

Kuhn's argument raises some important issues. One of the big questions I have addressed in this book concerns the degree to which our clumping the world into concepts and their relations affects our experiences of that world. The current question builds on this: to what degree does what-the-world-is depend on our experiences of that world? People who think that the world can get along just fine at being what-it-is without us, that reality isn't affected by what people believe about it, want to draw a distinction here between what-the-world-is and what-people-think-the-world-is, that is between *worlds* and *worldviews*. Their response is that Plutarch and I

7. Kuhn 1993.

describe different worldviews when we talk about magnets differently, but that there really is only one world, only one way things really are.

What Good Is Relativism?

My critique of certain realist arguments in the previous chapter does not, it should be emphasized, imperil every realist stance, although it does offer qualifications to many of the standard realist arguments. Of course, the historian may well feel at this point that the whole problem would be better solved by simply bracketing realism or by adopting some version of a relativist stance. Relativism, after all, has a lot to offer the historian.

The way some realists talk about relativism, though, one wonders how anyone could be so foolish as not to see the very hands waving in front of their eyes.[8] So what is the attraction? Many, but not all, versions of relativism come out of specific concerns about language and culture.[9] Many anthropologists have adopted versions of cultural relativism, for example, because their drive to understand other cultures without imposing or privileging the anthropologist's own categories, priorities, and ontologies—which has been shown to lead to all kinds of problems and distortions—forces them to think their way into whole other systems of understanding the world, systems that turn out to be conceptually consistent, complex, and empirically sound.[10] To see why an (inherently ethnocentric) realism tends to impede such an aim, one need only read some of the earlier thinkers whose conviction that we had it right and the objects of our imperious gaze did not, led them to treat their sources as incompletely developed, stuck, as it were, on the ladder that we looked down from the top of. If relativism is a clean way of avoiding attitudes that can only impede understanding of and between cultures, then it has a lot to recommend it. And not just in anthropology. Systematic cultural biases—biases that cultural relativism may help us to avoid—have led to well known fiascoes in everything from social policy to intelligence testing. Let us call this the utility argument for relativism.[11]

8. Charges of stupidity can be seen in the extreme dismissiveness with which some realists handle antirealists, where "absurdity" and questions of "sanity" are frequent substitutes for argument. See, e.g., the citations Devitt uses to support his own charge of the "prima-facie absurdity" of the "bizarre doctrine" of worldmaking (Devitt 2006, n. 7).

9. See, e.g., Douglas 1975.

10. Some recent attempts to move beyond a strict dichotomy of which relativism forms one pole can be found in, e.g., Descola 2004; Lloyd 2007.

11. See, e.g., Herskovits 1972. For a summary of the argument, see Fernandez 1990.

To be sure, one can find a few, usually outspoken, supporters of the "nothing is true, everything is permitted" school, but often the stance is taken more for the sake of shock value than anything, and more than one realist has suspected (with good reason) that, when push comes to shove, this position is often more metaphorical than metaphysical. And many critics suspect that too strong a relativism may have undesirable political and ethical consequences—must a relativist accept that early modern witch hunts were ethically justified? I think not, but to develop a consistent philosophical position that allows for some degree of cultural relativism without implying strong ethical relativism is not my concern here, so much as to point out that one can argue for the validity of actors' categories and contexts without committing to validating all actors' categories in all contexts.[12]

One persistent ethnographic problem to which relativism is an especially promising solution is the question that has been asked repeatedly in the latter half of this book: What do we do with the sometimes baffling beliefs we find in our sources? If we resist the move to relativism, then we have only a limited range of other responses, handily catalogued by Dan Sperber as follows:[13] (a) we can simply class our sources as acting irrationally; (b) we can dissolve this apparent irrationality into mistaken beliefs about the world grounded in incomplete information (note that both [a] and [b] may hide an ultimately unhelpful ethnocentrism, or from a historian of science's perspective, Whiggism); (c) we can take the *act of the statement* of irrational belief as primarily a semiotic performance, rooted in the complex interrelationships of culturally specific ideas, norms, and practices—which, for all it can usefully reveal about the ideas behind the utterance, has the unfortunate side effect of reducing all statements to what are essentially metaphors. To all of these, relativism looks a promising alternative. As the relativist sees it:

> It is not surprising, in particular, that the theoretical assumptions of another culture [e.g., the existence of a witchcraft substance or of spirit possession] should quite often seem irrational: such assumptions relate to actual observations through implicit inferential steps which it is easy for members of the culture and generally impossible for aliens to reconstruct. . . . Furthermore, it can be argued that the acceptability of propositions does not rest on observations and inference alone, but also on a

12. Hacking 1982 helps clarify this point by separating *relativism* from *subjectivism*.
13. Sperber 1982.

> number of *a priori* beliefs, or postulates. Such postulates determine a "world-view" within which the rationality of beliefs is to be assessed.[14]

As a methodological statement, this looks like it could be a very good candidate for having been one of the guiding principles of this book, and many readers will no doubt have suspected something like it was lurking in the corners. Indeed, I have gone to great pains to carefully elaborate in Roman sources exactly the points on which this argument hinges: observations, implicit (and explicit) inferential steps, postulates, cultures. . . . Moreover, the fact that the very criteria for rationality come from within the system itself and not from outside has been central to my position in this book. Nevertheless, for all of relativism's promise in the context of the problem of apparently irrational belief, the position I have adopted is not relativist, for very important reasons.

One is the worry about what is judiciously hidden in the ellipses in the above quotation, which is a lot of talk about translation and untranslatability. If the position as stated above is taken as correct, then in order to be consistent it seems that we cannot avoid also committing to some version of an untranslatability of experience and theory between cultures, or, to put it in Kuhnian terms, to incommensurability. This is because not only the standards for rationality, but also the meanings of terms, the interrelationships of concepts, and the permissible cognitive operations that we can perform on them, all of these are culturally specific. Dragging a term, an entity, a concept from one worldview into another involves a decontextualization that in turn means at best approximation and at worst misunderstanding when it is employed or explained in the other worldview. At the softer end, approximation, the historian or ethnologist can argue that long-term exposure to another culture can give one ways of seeing its terms, concepts, and categories that allow for at least proximate (and presumably also improvable) translation, and so over time I can come to find ways of understanding and explaining Galenic pneuma that do not necessarily imply that I have found some neutral third ground from which to view both Galen's world and my own and through which to shuttle material back and forth.[15] But precisely how—or even whether—we can make this position methodologically sound while avoiding a slide into a full-blown incommensurability or untranslatability is unclear.

14. Sperber 1982, 156. Note that Sperber does not in the end endorse this position.

15. A good argument leading to a compatible conclusion can be found in Lloyd 2004, chap. 7. See also Chang 2004; Lloyd 2007.

But there is a second problem of meaning, one that has to do with how it is that we come to understand the Roman cosmos in the first place. Think of the recurring entities in this book that modern science has no equivalent for, not even approximately: pneuma, sympathy, the gods. How is it that we come to have an idea of what they mean? We read books in which these terms occur, just as the ethnographer talks to people who use foreign concepts. Of course, dictionaries provide us with terms in our own language that we can, as a preliminary step, map the Roman terms onto, but that is only a beginning. A fuller understanding comes only with a fuller sense of the contexts in which the terms occur, the relationships these terms have with other terms, what the terms do under certain logical operations, and so on. All of this can be done more or less internally to the Roman worldview, that is, the more we think our way into their world and the longer we spend there, the richer—and clearer—the entities we are trying to understand become. I have used a range of analytic categories (embeddedness, symmetry, tropicality, etc.) to help fill in different parts of the puzzle and shed light on some of the more obscure connections. Throughout the book there has also been an invisible partner to our understanding, one we do not typically pay especial theoretical attention to but which has had a massive influence all the same: the world itself.

In trying to come to grips with the theories of vision in Ptolemy and Galen, for example, or in looking at sound in Seneca, we had recourse to a range of familiar entities, from walls to eyeballs, stars, nerves, paintings, and perfume. When Ptolemy talks about the way in which one type of column looks straight from a distance but tapered from below, we think we know what he is talking about—not by tracing some chain of literary references from antiquity to the present, but because we have seen columns out in the world. Similarly dogs and mirrors, maybe even earthquakes and volcanoes. All the theories we have been discussing in this book are theories about something, the world, that persists and whose observable behavior in the here and now is indispensable to our understanding of what ancient science *is.* Historical relativism, like its ethnographic counterpart, may be very useful for avoiding some of the bigger methodological problems we are faced with in theory change and differences in worldview, but the very agreement that we are talking about something, that we follow our informant's finger when he points at a rabbit and says *gavagai,* presupposes some thing out there on which there can be relativist perspectives in the first place.

Thus far, then, we are at least committed to a bald version of what is called metaphysical realism: that there is something out there, something real, that pushes back when we poke at it. But we must go a little farther still, beyond bare metaphysical realism. This is because the ways in which the world pushes back help us, here and now, to understand alternative ontologies and classifications. For this to work, it means that the possibilities for the world's action cannot be unlimited. There are a lot of ways of seeing lightning, to be sure, but coming to understand one perspective from another must always revolve around a shared phenomenal nexus. It is not just that in studying ancient science we are betraying a belief that *something* exists, but that something that has *properties very like those ascribed to it,* by both sides, exists. Our experiences and our understanding, then, are constrained by what the world is, or more specifically how it acts in response to our senses in light of the questions we put to it; and this fact is atemporal, applying equally to the Romans as to ourselves. But as we have seen again and again over the course of this book, preexisting taxonomies, preexisting ontologies, intellectual and educational contexts, all of these deeply condition how the phenomena we encounter in the world obtain meaning, how they make sense.

Coherence

One of the most useful epistemic guidelines available to the historian of science is a principle of what we might call *coherence charity*—that the way to understand how Roman authors make sense of the world is to pay attention to how their best descriptions of the world cohere, both with each other and with people's experiences of the world. They do not cohere exceptionlessly—that would be too much to ask of anyone's world—but they cohere very strongly for all that. Worlds are made not just of entities, but of *rich* entities, embedded entities, conceivable entities, classes, rules, possibilities, and more. Worlds hang together and things are thick. And it is just here that the sharp division between worldviews and worlds begins to blur. The world is an independent thing, but in any way we talk about it, it is always already the object of our experience. True enough, the horse would still be in the field munching its hay if we all vanished tomorrow, but there would be nothing to say about it.

On the face of it, a naïve coherentist epistemology—where knowledge claims are justified only by how they hold together within the larger framework of other knowledge claims—may look to offer an easy solution to the problem of difference between the world of the Romans and our own. The Romans had a set of coherent beliefs; we have a different set; and each of us is justified in our own web. Coherentism differs from epistemic foundationalism in not needing to specify some final foundation on which the whole edifice builds. Otto Neurath offers one of the key metaphors for coherentism when he says that we are like people at sea on a boat. As we notice problems we make repairs and contrive patches to keep the boat afloat, working bit by bit, but there is no shore to which we can retreat to rebuild the ship from scratch.[16]

But we might be fools to rush in too hastily: by emphasizing coherence as a criterion, where good ideas are the ones that cohere better with all our other ideas, we face a potential Trojan horse. Our commitment to coherence may mean that the truth-values of someone's beliefs may depend solely on the *interrelations* between their beliefs, with no recourse to the relations between their beliefs and phenomena in the natural world except with regard to what they already believe about those phenomena.[17] This creates two immediate lines of objection, one philosophical, one historiographic. A philosophical concern many would voice here is that if truth rests merely in coherence between ideas rather than in some kind of agreement between what I believe and the world-out-there, then there are multiple truths—and if multiple, then none. We run to scepticism via extreme relativism. Moreover, it would seem to be difficult or even impossible to give an account of new knowledge acquisition, let alone an account of how it is that the world doesn't always behave as expected. The world, in short, seems to hit back sometimes. Finally, there is a common philosophical worry about circularity, if belief acceptance hinges only on agreement with what is previously accepted. Fortunately, good versions of coherence avoid both of these worries.

The historiographic worry is that on a coherentist view, the set of coherent truths held by the Romans differs from the set held by us. Not only are the theoretical entities different, but so are the laws about how entities interact, the classifications of entities, the rules and conventions for analyzing and adjudicating, and even, as we've seen, empirical phenomena. But, says the coherentist theory of truth, that's acceptable because they have their truth and we have ours, and the only real criterion is how it all fits together. This idea has the same cultural benefits that we saw relativism to have, but it also has the same costs (indeed, this position is just one particular stripe of relativism). On the other hand, many realists want to argue that some approaches to the natural world really are better. Science, for ex-

16. His original statement is Neurath 1932–33: "Wie Schiffer sind wir, die ihr Schiff auf offener See umbauen müssen, ohne es jemals in einem Dock zerlegen und aus besten Bestandteilen neu errichten zu können."

17. See, e.g., Haack 1993, chap. 3; BonJour 1985, 108f.

ample, gives us antibiotics, improved crop yields, the internal combustion engine, household electricity, Doritos. Although most of us would acknowledge that there are downsides to all of these as well, few of us have chosen to forego the lot of it on principle.

There is a third problem that blurs the boundaries between philosophy and historiography. If coherence is the way both to explain and to validate the differences between the world of the Romans and our own, then we seem to have demarcated two belief-bubbles with no possibility of content overlap, no real world out there that we are both talking about, no third source for the historian to appeal to as she tries to come to grips with Seneca's theory of lightning. The way to get around this is by drawing a clearer distinction between coherentist accounts of *truth* and coherentist accounts of *justification.*[18] In the coherentist account of justification, we do not say that the truths of the Romans were true because they cohered in a belief-bubble—for that would be a coherentist account of *truth*—but instead that the Romans were *justified* in believing what they did because what they believed was coherent, both theoretically and empirically. This second claim has not yet made any appeal to truth, only to justification of belief. Still, if coherence is all I have as a criterion for the justification of belief, is not coherence my ultimate arbiter of truth as well? Certainly this is taken as a common concomitant, one often enough inferred by proponents and critics of coherentism alike, and one that more or less directly runs us into some version of relativism and incommensurability.

But not all justification-coherentists are truth-coherentists. There are several good arguments for combining coherence theories of justification with good old-fashioned correspondence theories of truth, where sentences are true if and only if they *correspond* to how the world really is.[19] Donald Davidson offers a different argument, one that he initially thought led from coherence justification to correspondence truth. On later reflection, though, he argued that the version of truth he was calling correspondence in that paper was not in fact correspondence but, broadly speaking, pragmatist.[20] We will return to Davidson's actual argument shortly, right after we clarify

18. For the importance and implications of this distinction, see Thagard 2000, chap. 4; BonJour 1985, chap. 5.

19. See BonJour 1985, chap. 8; Thagard 2000, 78f., 85f.; Kuukkanen 2007; Haack 1993, comes to some version of truth that is non-coherence-based, though I am not positive whether it is correspondence-based or pragmatist. I suspect correspondence only because of the way she limits veridicality to truth-indicativeness rather than to truth itself, where talk of the latter would not seem to be strictly precluded by most pragmatist theories of truth.

20. Davidson 1983, 2001.

what a pragmatist theory of truth looks like, and what its advantages and disadvantages are in light of the historical challenges we are faced with in the ancient sciences.

Truth and Meaning

History challenges many of our accepted theories of truth, and it is not entirely clear where the solution lies.[21] I propose in what follows to argue that some version of a pragmatic theory of truth best meets the challenges posed by historical sciences and historical realisms alike, but I can do no more than gesture in a promising direction. What are the available options? Correspondence theories of truth, the most commonly championed type of truth theory, claim that true ideas are ones that correspond with the way the world is. Familiar problems with correspondence theories arise around what we mean by "correspond," and how or whether we could ever know that our ideas were so corresponding.[22]

Pragmatist theories of truth, on the other hand, pay close attention to how we come to believe something is true, and to how we verify and test that belief in experience. In the classic formulation of William James—a formulation that is not unchallenged even among pragmatists but is still a useful starting point—we get:

> *True ideas are those that we can assimilate, validate, corroborate and verify. False ideas are those that we cannot.* . . . The truth of an idea is not a stagnant property inherent in it. Truth *happens* to an idea. It *becomes* true, is *made* true by events. Its verity *is* in fact an event, a process: the process namely of its verifying itself, its veri-*fication*.[23]

Given the earlier discussion, we should also note the criterion of assimilability in the above definition, which means for James that truth needs to be, as we might say, coherent as well as verifiable. James also handily

21. See Krüger 1991.

22. A more subtle line of objection can be found in Lewis 2001.

23. James 1907, 97, emphasis his; "veri-*fication*" [*sic*]. There are other theories of truth that get called pragmatist, but I think they will not be useful for us here. E.g., Charles Peirce's definition of truth ties truth to the outcome of a hypothetically completed inquiry, which for our purposes would mean that physical truths are what we would believe when physics is finally finished. I simply have no way of understanding what it would mean for physics to be finished. Nor can I see how deferring truth to a hypothetical time in the future helps us clarify what truth means (compare James' complaint at James 1907, 106; and Rorty 1991, 131).

carves out explicit room for testimony: "You accept my verification of one thing, I yours of another. We trade on each other's truth. But beliefs verified concretely by *somebody* are the posts of the whole superstructure."[24] So far, James' theory of truth has the advantage for our purposes that it looks to meet most of the challenges thrown up in this book, but the reader may suspect that a relativist trap is lurking just around the corner. For if truth is merely what coheres and has been verified in experience, then are we not in a situation such that the Romans had one set of truths and we another? Does anything go, so long as it works? Emphatically not. Although a Jamesian theory would seem to want to conclude that pneuma and sympathy were once true, the fact that they are no longer true is not just a function of our adoption of a different worldview, but instead a function of the nontrivial fact that in the intervening 2,000 years we think we have had experiences that disprove them or render them useless or superfluous. Magnetism has built around itself a host of tests and theories that were not part of the verification process available to the Romans, and so magnets-as-sympathetic have been shown to be false. This fact may also allow for a (qualified) notion of something like epistemic progress, without the normally attendant triumphalism.[25]

But if, for the pragmatist, verifiability leads to variability, what kind of *truth* is that? The pragmatist answer is: the only kind we have. Truth is a stamp of approval that we give things as a way of saying that they meet certain—and most rigorous—criteria. For the historian, a pragmatist definition of truth fits very well with a coherence-based epistemology that allows us readily to be realists about atoms and electricity, and at the same time to be good historiographers. The historian should be careful, though, not to see *time* as the primary variable between worldviews. Truths vary not just over time, but across cultures—and no, we do not need to be relativists to say that. Truths vary across cultures because the meanings of terms vary across cultures.[26] When an Amazonian says that jaguars are spiritually humans enveloped in a different bodily form, if I take him to actually be saying something meaningful about the world, then he must mean something very different by *jaguar* and *envelop* and *spirit* and *human* than I do.[27] Is it true that jaguars are humans in his sense? Maybe, maybe not. It depends

24. James 1907, 100, emphasis his.

25. See Chang 2004, 2009.

26. What follows is indebted to (but not identical with) Davidson 1983, and in some respects also to Strawson 1997.

27. I owe the example to Lloyd's reading of Viveiros de Castro 1998. See Lloyd 2007.

on whether the claim is verifiable, and whether it is verifiable depends utterly on what he means by it. If I can march in and demonstrate that there is some insuperable contradiction in his categories, or that there is powerful empirical evidence that, given what he means by the terms, what he says about them cannot be true, then both he and I are compelled to admit that his claim was false. (Something precisely analogous to this is what happens when theories and classes and ontologies come to be rejected historically.) If, on the other hand, he is able to convince me that, given his meanings of the words, the jaguar really is a human spirit in cat's clothing, then I have to confess that he is speaking the truth. What separates this from relativism is the emphasis on meaning, meaning clarification, and intertheoretical discourse. It's not that anything we can say about the world must be true within the framework in which we say it, but that there are a lot of things we can say about the world—many different ways of talking about it—and the judgment of whether any one way of talking about it is true rests in whether it bears up to the most careful scrutiny.

Realism, Coherence, and History

One legitimate complaint about coherence-based epistemologies is that they don't have a way of privileging experience over already-held belief, so we cannot get to anything like the pragmatist theory of truth from a coherentist epistemology (or vice versa). Davidson offers a clean way of fixing this problem by beginning with the observation that for most people, most of their beliefs about the world work pretty well. So, too, most of their beliefs about the world cohere with each other pretty well. Asking the sceptical question about how we know any one belief (alternatively: each and every individual belief) to be true is simply to miss the boat, insofar as the very possibility of communication presumes that this "mostly working pretty well most of the time" should presuppose that most of our beliefs are true, and that any one belief, because it coheres with the rest and the rest works, should be presumed true under most circumstances as well. Of course, the sceptic could urge that philosophy is not "most circumstances," but the sceptic's demand that we start from some one unshakable grounding point, one foundation, misunderstands how belief and truth are established (a point I have argued repeatedly in this book). In order to decide whether what the Amazonian is saying about jaguars could be true or not, I must first understand what he is saying, but I do not come to understand what he is saying by asking him to ground every proposition he makes in some indubitable way, nor by finding some neutral third ground from which to bring his ideas into relation with my own.[28] Instead I try to clarify this bit over here, to explain that bit over there, to relate this third bit back to what he said earlier about something else. Slowly it starts to make sense, but it makes sense as a whole. Likewise when we read Cicero and try to see how he can say one thing about divination in one text and then apparently say the opposite in a second text, we do not come to understand what he means by running each and every sentence of his back to individual and precisely demarcatable foundational claims, but by reading our way into a whole rich world in which his sentences relate to each other and so make sense, logically, socially, taxonomically, ontologically.

Here Davidson draws our attention, as I did above, out to the world itself as an important actor in how we understand someone:

> It is impossible for an interpreter to understand a speaker and at the same time discover the speaker to be largely wrong about the world. For the interpreter interprets sentences held true (which is not to be distinguished from attributing beliefs) according to the events and objects in the outside world that cause the sentence to be held true.[29]

As we are reading or listening to someone, we take their sentences to be about something (in this book, about things in the world), something that has caused them to believe what they do. We think that we understand what they are saying about that something when we see how that something causes the belief that we ascribe to them. Seeing that link is seeing what the belief-statement *means*, which, for Davidson, is the same as seeing how it is true. Of course, this won't work in every instance (nor for every type of sentence), but if we presume, as Davidson did at the outset, that most of what people believe about the world is true, then it will usually work, and it will work most often for the most central and most important sentences. This theory of truth isn't precisely James', to be sure, but it is related in many respects.[30] One important feature that it shares with James, as with my analysis above, is an agreement on the significance of our experiences of the world to how we understand each other.

28. This last point takes the wind out of one common relativist argument, which sees the lack of a third ground as necessitating a shift to relativism. See Lloyd 2007, 3.

29. Davidson 2001, 150.

30. Davidson wants to distance himself from most pragmatist theories of truth, but at the same time acknowledges the important resemblances with his own account. I agree with Rorty: Davidson's is *a* pragmatist theory, not *the* pragmatist theory (see the *Afterthoughts* to Davidson 2001).

The question of exactly how much weight to give experience in coherence is a difficult one, partly because coherence is of necessity an extraordinarily complex set of criteria. Indeed, it is its complexity that has led many of its critics to despair of finding any finite means of adjudicating between competing knowledge claims from within a coherentist epistemology, since it is not a trivial matter to determine how—or even whether—we can measure the degree of coherence one claim has with a given theory, versus another claim.[31] Nevertheless, careful differentiation of different types of coherence, and careful attention to their relative weights and constraints, has shown not only how coherence determination is in fact tractable, but also how coherence itself models the ways in which reasoning takes place out in the real world, both philosophically and psychologically (and the psychological aspects are appealing, as not all of the competitors to coherence offer this level of explanation).

For our purposes, one useful aspect of Paul Thagard's analysis is his division of coherence into subtypes: explanatory coherence (when a hypothesis explains, by providing a causal account of a phenomenon); analogical coherence (when a hypothesis coheres with a body of theory by being analogous to some accepted part of that body, as when Darwin drew analogies between the not-yet-accepted mechanism of natural selection and the accepted mechanism of artificial selection, or when Seneca drew our attention to clapping hands); deductive coherence (when a hypothesis coheres because it is deduced from accepted theories, or they are deducible from it); perceptual coherence (when we interpret visual stimuli as meaning something about place, size, shape, and type of object); and conceptual coherence (when concepts fit together meaningfully).[32] I have not, in this book, deliberately cut up the pie into neat Thagardian segments, but it is no great stretch to pick apart many of my arguments and to reassemble them into these categories: deductive (ubiquitous and of significant import to our sources), analogical (modeling, classification), perceptual (embedded seeing), and conceptual (narrowly understood, classification, but widely understood, ubiquitous)—

31. Solutions have run along three lines: (1) being satisfied with nonquantifiable accounts of coherence (BonJour 1985; Haack 1993; and perhaps also Davidson 1983), (2) trying statistical methods with what look to me to be unrealistically small data sets (see, e.g., Stoneham 2007 and his respondents Jäger 2007 and Rowbottom 2007; Siebel and Wolff 2008; compare also Shogenji 1999; Bovens and Olsson 2000; Olsson and Shogenji 2004), and (3) Thagard's (1992, 2000) artificial-intelligence modeling, which uses a computer program called ECHO to calculate complex (and rather odd-looking) decision trees.

32. See Thagard 2000.

all these types of coherence have had considerable play in our analysis of Roman science.

The appeal of finding a single methodology for understanding both electrons and dogs in my experience and for understanding pneuma and nerves in Galen's, is obvious. We must be careful, though, not to say that what I find philosophically satisfying in coherence would have been philosophically satisfying to Galen. We can, however, still pay close attention to the psychological importance of coherence for theory choice, an importance that finds significant historical confirmation in this book. I argue as a corollary that a historiography that attends to the multiple and rich coherence aspects of theory choice gives us the fullest answer possible to the question of how the Roman sciences make sense. The answer: only as a complex and embedded whole. What do the sciences make sense of? The world; and that world is *real.* The corollary question about why that world sometimes looks so odd turns out to have a lot to tell us about how we understand the world as well.

But that real world, when described, when understood, when talked about, is represented in languages, and the statements we make about it may be true or may be false. Taking a pragmatic definition of truth allows us to maintain a consistent realist stance but emphasizes that truth is something that happens to statements when they meet experience. As was the case for Ptolemy, since no one has had all possible or all future experiences, we cannot know in the here and now what experiences will meet our statements in the future or will meet future versions of our statements. "True," then, is subject to change. Does this not amount to relativism? I say not, most immediately because of the conviction that the world I am experiencing, the keys that I am typing, the audience that I am addressing, the electrons spraying the back of my computer monitor, all of these really exist, and really exist more or less as I conceive them to be. Not only are they real, but I—we—have a pretty good picture of them (not to mention a remarkable and impressive ability to manipulate them). But we must also not forget that I cannot now reasonably adopt a sympathetic or pneumatic worldview, simply because too much has changed in the intervening millennia. To take just one aspect: the data set is not the same. Magnets have been shown to do things that sympathy cannot and was never intended to explain. Our world is not just a different way of looking at their world, but is instead something completely new and unanticipated by them.

Here Brian Ellis gets us back to a polyvalent realism: "If the world behaves as if entities of the kind postulated by science exist, then the best explanation of this fact is that they really do exist."[33] I think this hits the nail on the head exactly, even if the historical view changes considerably how we read it. The world behaves *as if* entities of the kind postulated by good science exist *and it always has behaved that way,* even when the entities that good science postulated were very different.[34] The argument of this book shows that the realist needs to recognize that, historical contingencies being a little different, I—we—would just as firmly have been realists about sympathy as we now are about electrons. That sympathy/antipathy is no longer something we can be realists about is simply—*only*—the consequence of a long, strange, and unique historical journey.

In this sense, then, what I called the virtues of relativism in this chapter may perhaps be better parsed as the virtues of what is called by philosophers "epistemic contextualism," where knowledge claims are to be adjudicated within the contexts in which they are uttered.[35] Thus given a statement like "I know that I am sitting at my desk," what is known can be said to be known only within an appropriate epistemic context, one in which sceptical questions involving Descartes' demon or *The Matrix* have no place. By extension, I want to say that pneuma and antipathy were known in a (historical) epistemic context, even if the assertion of them today can no longer be taken as a viable knowledge claim. Time, in this case, becomes an important epistemic context-variable. As in classic versions of epistemic contextualism, we still allow room for the sceptic or the curious onlooker to interject some sobriety or clarification when necessary—we can acknowledge that when the context is shifted by asking some additional question like "how do you know you are not dreaming?" or, "is that surprising claim in Ptolemy really true?"[36] we can give answers that commit us neither to blanket scepticism nor to blanket Whiggism. At the same time, epistemic context allows us to see the full force of two of the key words in the book's title: What *did* the Romans *know*?

33. Ellis 1988,413.

34. Indeed, this very point must be acknowledged even by hard-core realists like Devitt 2006, insofar as such experiences form the starting point for his argument for realism, which begins with the observation that—outside of a limited set of debates in academic circles—everyone thinks that the things they believe in are real, and that the impression of reality is confirmed by experience day by day. Devitt does not unpack whom he means by "everybody," but if his argument is to hold it must exclude "everybody who does not believe what I do," if he is not to grant Galen a realism as well.

35. See, e.g., Annis 1978; Lewis 1996; Stanley 2005; Preyer and Peter 2005.

36. By which I mean, did he see something that we (or I) were simply not aware of but that, when checked, turns out to be observable?

CHAPTER ELEVEN

Conclusion

I end this study as an epistemological coherentist. This should occasion no surprise, insofar as I emphasized from the very beginning the ways in which sense is made of the world only through a wide range of conceptual and epistemological interconnections. I recognize that the correspondence theory of truth is incompatible with the historical and epistemological picture I have been arguing for. Finding himself in an analogous position, Hasok Chang chose to circumvent the problem by "looking away from truth" more or less entirely, turning instead to a self-contained notion of epistemic progress that is independent of any notion of what-is-true, and consists instead in the improvement of certain "epistemic virtues," things like elegance, accuracy, testability, generality, and so forth.[1] An improvement in the accuracy by which we measure a phenomenon is then a perfectly acceptable example of epistemic progress just by virtue of being such an improvement, and we need no recourse to any notion of having come any closer to something called "the truth." This is certainly a sensible and consistent reply to the problem that truth-correspondence poses for him and myself alike. But I think we can do a little more for truth than simply turn away from it. There is after all one notion of truth available off the philosophical shelf that would still satisfy Chang's criteria, although it is not a truth that one can come "closer to," even if it does allow for revision. This is the pragmatic theory of truth, which says that we count as true any propositions that have passed the best tests we can think of to throw at them.

That being said, though, coherentism as an epistemological position is not usually elaborated with respect to just the kinds of problems I have used it to solve in this study. I tried to show how coherentism wedded with

1. Chang 2004, 224f.

pragmatist truth could take advantage of some of the promise that relativism seems, prima facie, to have for historiography, while allowing historians to face an implicit realism that lies at the heart of how they come to understand ancient sources, a realism that happens also to underlie my own trust in computers, vaccines, and airplanes. The important caveat that one cannot *now* rationally just switch to a Roman worldview as though it were an option among many others, emerged when we considered the profound changes that have taken place in the intervening millennia with regard to what we know about the world and how we know it—the historian's ability to code-switch notwithstanding. I offered a (very preliminary) reading of that code-switching in terms of epistemic contexts at the end of the last chapter, but again, philosophers have not elaborated contextualism in order to solve the problems of historical sciences, so much work remains to be done to see how this conjunction of philosophical strategies will play out in the broader historiography of the sciences.

All of this emerges both from an initial desire to give Roman investigations into nature their due as sophisticated and interesting epistemic projects, and from some of the conclusions that came out of that investigation, particularly those arising from knowledge claims in our sources that centered on things we no longer find compelling or even plausible, whether divinities, divination, or antipathy. The genealogy of the project as a whole may be telling here: it began with my utter surprise and delight at finding the now familiar framing of garlic-magnet antipathy in Plutarch. That surprise generated what became chapter 6, and the rest of the book emerged as attempts to tackle or tighten up various parts of the same epistemic problem set, all centered on how things hang inextricably together to make up a world, how change over time can be understood without being Whiggish or relativist, and on how very complex and interdependent systems of knowledge were and continue to be.

This interdependence is not limited to an interdependence of what we might loosely call scientific theories or observations, but further includes a wide range of factors (many of them culturally specific), in the Roman case including most prominently theology and law. Not only did divine providence filter down, for many of our sources, into an implicit order and rationality inherent in the cosmos itself, but it also underscored moral theories rooted in nature. Morality paid back this debt to nature by allowing Seneca to use individual ethical virtue as a guarantee of reliability for otherwise problematic observation claims made by witnesses of novel phenomena. Investigating how those and other phenomena were incorporated into theories brought us into a series of other multidirectional relationships, between seeing and knowing, between physics and free will, between numbers and the soul, between heaven and earth themselves.

In several chapters, I emphasized the ways in which judgments on the acceptability of individual explanations *as explanations* must be seen as contextualized within a discourse community, such that "like affects like," cosmic harmonics, and number theory could be offered as satisfying ends to chains of why-questions. We also touched repeatedly on the contexts of politics, education, performance, and intertheoretical criticism and debate, in order to see how broadly ways of thinking about the world reverberate, playing back and forth.

If garlic and magnets began this project, perhaps garlic and magnets should end it by broadening antipathy out as a metaphor for the set of problems at the heart of the book as a whole. The tension that emerges in the space between my immediate reaction to Plutarch's claim and his certainty in its truth cannot be simply dispelled with an empirical test—there are far too many other examples out there for this to be practicable or even meaningful for every and all cases. Instead, sooner or later, I have to trust the larger web of my own knowledge about the world, its classifications, its ontologies, its theories and observations, any one of which may be fallible, but the whole of which works together to paint a reliable, effective, and useful picture. I realize that this is just the position Plutarch himself had been in with respect to the question of how magnets worked, but my knowing now something different than he knew then relies on my position at the end of a long temporal trajectory. I hope the metaphor will not be too far a stretch, but I cannot resist quoting my favorite version of one Heraclitus fragment here: "This river I step in is not the river I stand in."[2]

2. Inscribed, for reasons that entirely escape me, on the Queen Street viaduct in Toronto.

APPENDIX

Lemma to the Mirror Problem

Not everyone is immediately happy that the mirror problem has a straightforward solution. Standard answers run the problem back to the psychology of bilateral symmetry and the relativity of left and right as directions versus up and down as absolutes,[1] but as the physicist Brendan Quine forced me to see, this is not quite the whole story. Quine's objection runs as follows: Imagine you are looking in a mirror, holding an apple in your right hand. Now imagine someone standing where the mirror is and photographing you. When you look at the photograph taken from the mirror's point of view, the apple will be in your right hand in the photo, which is (in an absolute sense) on the opposite side to where the apple appears in the mirror, as in figure A.1.

The apple appears to be in the wrong hand in the mirror, but when it appears in the right hand in the photograph, it has also switched sides in an absolute sense relative to the mirror image. In order to reproduce the mirror image, we need to turn the photograph 180° horizontally and look through it (imagine holding it up to the light, or that it is a photographic slide). What we do not do, Quine points out, is turn the photograph 180° vertically. Does this not mean that there is something more going on than just "it looks like it is in the wrong hand only because you project right and left into the mirror image?" Are up and down not more absolute here?

My answer is this: yes and no. The real villain in the piece has a good deal to do with how we move around in the world, *viz.*, on a horizontal plane. The reason the photograph needs to be turned horizontally to reproduce mirror conditions has to do with how the photographer oriented herself relative to the viewer in the first place, and accordingly how we develop and then hold photographs to reproduce that image "correctly." Imagine the photographer initially

1. See Block 1974; Denyer 1994.

Fig. A.1 Switching sides only relatively, or is there something absolute going on? Why is the vertical axis privileged so that we have to rotate a photograph around it in order to re-create mirror conditions?

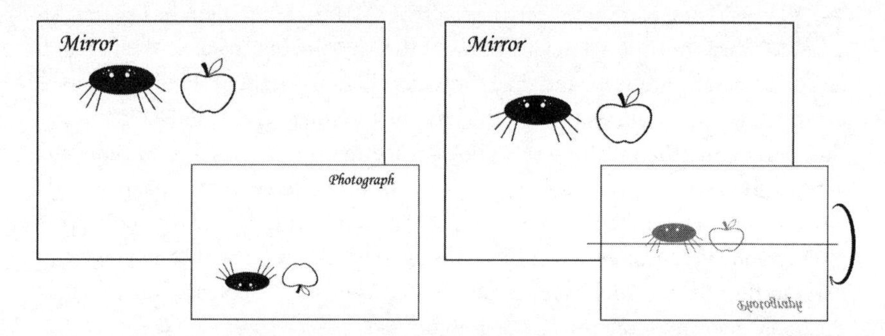

Fig. A.2 Because of their different axis of rotation, spiders in a beach ball would hold the photograph differently in order to reproduce the conditions of initial observation. Thus left and right become absolute, and up and down relative. In this instance, the mirror puzzle must be reworded: Why do up and down get reversed in a mirror, but left and right do not?

standing beside you as you prepared the experiment. Before taking the photo, she had to stand between you and the mirror *and turn around 180° horizontally* in order to see you and the apple. It is this process that the horizontal inverting of the photograph is meant to compensate for. Had we been, say, spiders living on the inside surface of a beach ball, the situation could have been completely different, insofar as the photographer spider could have (or likely would have) walked to the opposite side of the beach ball by going up along a meridian rather than around the equator, so to speak. When she arrived opposite you, her right side and your right side would both be absolutely in the same direction (pointing east, say) but her head would be pointing down and yours up. In such a world it would be natural to hold photographs upside-down rather than as we do, right-side-left, in order to reproduce the conditions of initial observation. (I suppose that this would also apply to flagellum-propelled animals in open water, among other creatures real or imagined.) Thus left and right become the absolutes and up and down are inverted, and the mirror problem requires a reciprocal wording to our version. Finally, in order to produce the mirror image, we have to turn the photograph 180° vertically and project through it, rather than 180° horizontally as we needed to do in the human case (see fig. A.2).

REFERENCE LIST

d'Abano, P. 1476. *Conciliator differentiarum philosophorum et medicorum*. Venice.

Abry, J.-H. 1988. "Auguste: La balance et le capricorne." *Revue des études latins* 66:103–21.

———. 1993. "Manilius et Germanicus." *Revue des études latins* 71:179–202.

———. 1996. "L'horoscope du Rome." In B. Bakhouche, A. Moreau, and J.-C. Turpin, eds., *Les astres: Actes du colloque international de Montpellier, 23–25 mars 1995, Séminaire d'étude des mentalités antiques* 2:121–40.

Achinstein, P. 1994. "Explanation vs. Prediction: Which Carries More Weight?" *Proceedings of the Philosophy of Science Association* 2:156–64.

———. 2002. "Is There a Valid Experimental Argument for Scientific Realism?" *Journal of Philosophy* 99:470–95.

Algra, K. et al., eds. 1999. *The Cambridge History of Hellenistic Philosophy*. Cambridge.

Allen, J. 2001. *Inference from Signs: Ancient Debates about the Nature of Evidence*. Oxford.

Amouretti, M.-C. 1998. "La transmission des connaissances agronomiques dans l'antiquité." In M.-C. Amouretti and F. Sigault eds., *Traditions agronomiques Européenes*. Paris, 17–26.

Amouretti, M.-C., and G. Comet, eds. 1995. *La transmission des connaissances techniques*. Aix-en-Provence.

Anderson, G. 1993. *The Second Sophistic: A Cultural Phenomenon in the Roman Empire*. New York.

Ando, C. 2008. *The Matter of the Gods*. Berkeley.

Annas, J., and J. Barnes. 1985. *The Modes of Scepticism*. Cambridge.

Annis, D. B. 1978. "A Contextualist Theory of Epistemic Justification." *American Philosophical Quarterly* 15:213–19.

Armisen-Marchetti, M. 2001. *Macrobe: Commentaire au Songe du Scipion*. Paris.

Armstrong, D. 1978. *A Theory of Universals*. Cambridge.

———. 1983. *What Is a Law of Nature?* Cambridge.

Asmis, E. 2008. "Lucretius' New World Order: Making a Pact with Nature." *Classical Quarterly* 58:141–57.

Asper, M. 2007. *Griechische Wissenschaftstexte*. Stuttgart.

Audi, R. 1997. "The Place of Testimony in the Fabric of Knowledge and Justification." *American Philosophical Quarterly* 34:405–22.

Ax, W. 1938. *M. Tullius Cicero: De divinatione, De fato, Timaeus.* Leipzig.

Baader, G., and R. Winau, eds. 1989. *Die hippokratischen Epidemien: Theorie, Praxis, Tradition.* Berlin.

Bachelard, G. 1934. *Le nouvel esprit scientifique.* Paris.

Barchiesi, A. 2005. "Centre and Periphery." In S. Harrison, ed., *A Companion to Latin Literature.* Oxford, 394–405.

Barker, A. 1989. *Greek Musical Writings II.* Cambridge.

———. 2000. *Scientific Method in Ptolemy's "Harmonics."* Cambridge.

Barnes, J., ed. 1994. *Aristotle's Posterior Analytics.* 2nd ed. Oxford.

———. 1997. *Logic and the Imperial Stoa.* Leiden.

Barnes, J., and M. Griffin, eds. 1997. *Philosophia togata II.* Oxford.

Barnes, J., and J. Jouanna, eds. 2003. *Galien et la philosophie.* Geneva.

Bartholomaeus Anglicus. 1495. *Bartholomaeus de proprietatibus rerum.* Westminster.

Barton, T. 1994a. *Power and Knowledge: Astrology, Medicine and Physiognomics under the Roman Empire.* Ann Arbor.

———. 1994b. *Ancient Astrology.* New York.

———. 1995. "Augustus and Capricorn: Astrological Polyvalency and Imperial Rhetoric." *Journal of Roman Studies* 85:33–51.

Bartsch, S. 2006. *The Mirror of the Self.* Chicago.

Bazin, A. 1975 [1945]. "Ontologie de l'image photographique." In *Qu'est-ce que le cinéma?* Paris, 9–17.

Beagon, M. 1992. *Roman Nature.* Oxford.

———. 2005. *The Elder Pliny on the Human Animal: Natural History Book VII.* Oxford.

Beard, M. 1986. "Cicero and Divination: The Formation of a Latin Discourse." *Journal of Roman Studies* 76:33–46.

———. 1993. "Looking (Harder) for Roman Myth." In F. Graf, ed., *Mythos in mythenloser Gesellschaft: Das Paradigma Roms.* Stuttgart, 44–64.

Beard, M., J. North, and S. Price. 1998. *Religions of Rome.* Cambridge.

Beatty, J. 1997. "Why Do Biologists Argue Like They Do?" *Philosophy of Science* 64:S432–43.

Beck, R. 2007. *A Brief History of Ancient Astrology.* Oxford.

Beech, M. 1988. "Meteor Astronomy: A Mature Science?" *Earth, Moon, and Planets* 43:187–94.

Bennett, D. 1969. "Essential Properties." *Journal of Philosophy* 66:487–99.

Bernard, A. 2003. "Sophistic Aspects of Pappus' Collection." *Archive for History of Exact Sciences* 57:93–150.

Berno, F. R. 2003. *Lo specchio, il vizio e la virtù.* Bologna.

Berthellot, M., and C. E. Ruelle. 1888. *Collection des anciens alchimistes grecs.* Paris.

Berryman, S. 1997. "*Horror vacui* in the Third Century B.C.E.: When Is a Theory Not a Theory?" In R. Sorabji, ed., *Aristotle and After.* London, 147–57.

———. 2009. *The Mechanical Hypothesis in Ancient Greek Natural Philosophy.* Cambridge.

———. Forthcoming a. "Rainbows, Mirrors, and Light: Can Aristotle's Theory of Vision Be Saved?" In M. Martin and M. Stone, eds., *Problems of Perception and Vision*. London.

———. Forthcoming b. "The Evidence for Strato of Lampsacus in Hero of Alexandria's *Pneumatica*." In W. W. Fortenbaugh, ed., *Strato of Lampsacus*. New Brunswick, NJ.

Bilde, P., ed. 1993. *Centre and Periphery in the Hellenistic World*. Aarhus.

Block, N. 1974. "Why Do Mirrors Reverse Right/Left but Not Up/Down?" *Journal of Philosophy* 71:259–77.

Bloomer, W. M. 1997. "A Preface to the History of Declamation." In Habinek and Schiesaro 1997, 199–215.

de Boate, A. 1653. *A rare and excellent Discourse of Minerals, Stones, Gums, and Rosins; with the vertues and use thereof, By D. B., Gent.* Appended to Plat 1653.

Bobzien, S. 1998. *Determinism and Freedom in Stoic Philosophy*. Oxford.

Bonadeo, A. 2004. *Iride: Un arco tra mito e natura*. Florence.

BonJour, L. 1985. *The Structure of Empirical Knowledge*. Cambridge, MA.

Bonner, S. F. 1977. *Education in Ancient Rome*. Berkeley.

Borg, B. E., and E. Borg, eds. 2004. *Paideia: The World of the Second Sophistic*. Berlin.

Bouché-Leclercq, A. 1899. *L'astrologie grecque*. Paris.

Bovens, L., and E. J. Olsson. 2000. "Coherentism, Reliability and Bayesian Networks." *Mind* 109:685–719.

Bowden, H. 2005. *Classical Athens and the Delphic Oracle: Divination and Democracy*. Cambridge.

Bowen, A. 2007. "The Demarcation of Physical Theory and Astronomy by Geminus and Ptolemy." *Perspectives on Science* 15:327–58.

Bowersock, G. 1969. *Greek Sophists in the Roman Empire*. Oxford.

———., ed., 1974. *Approaches to the Second Sophistic*. University Park, PA.

Boylan, M. 2007. "Galen: On Blood, the Pulse, and the Arteries." *Journal for the History of Biology* 40:207–30.

Brandon, R. 1997. "Does Biology Have Laws?" *Philosophy of Science* 64:S444–57.

Brewer, E. C. 1898. *Dictionary of Phrase and Fable*. Philadelphia.

Brewer, W. F., and B. L. Lambert. 2001. "The Theory-Ladenness of Observation and the Theory-Ladenness of the Rest of the Scientific Process." *Philosophy of Science* 68:S176–86.

Brody, B. 1973. "Why Settle for Anything Less Than Good Old-Fashioned Aristotelian Essentialism?" *Noûs* 7:351–65.

Brook, D. 1983. "Painting, Photography, and Representation." *Journal of Aesthetics and Art Criticism* 42:171–80.

Browne, Sir T. 1646. *Pseudodoxia epidemica, or, Enquiries into Very many received tenents, and commonly presumed truths*. London.

Brush, S. G. 1989. "Prediction and Theory Evaluation." *Science* 246: 1124–29.

———. 1994. "Dynamics of Theory Change." *Proceedings of the Biennial Meeting of the Philosophy of Science Association, vol. 2*, 133–45.

Burkert, W. 1962. *Weisheit und Wissenschaft; Studien zu Pythagoras, Philolaus, und Platon*. Nuremberg.

Cambiano, G. 1999. "Philosophy, Science and Medicine." In Algra et al. 1999, 585–616.

Canali de Rossi, F. 2001. *Il ruolo dei patroni nelle relazioni politiche fra il mondo greco e Roma in età repubblicana ed augustea*. Munich.

Cardano, G. 1582. *Hieronymi Cardani mediolanensis medici, De subtilitate libri XXI*. Basel.

Cardona, G. 1975. "Tolomeo e Galeno." In M. del Pra, ed., *La filosofia ellenistica e la patristica christiana dal III secolo a.C. al V secolo d.C.* Milan, 229–49.

Carman, C. C. 2005. "'Realismo científico' se dice de muchas maneras, al menos de 1111." *Scientiae studia, São Paulo* 3:43–64.

Carnap, R. 1966. *Philosophical Foundations of Physics*. New York.

Carroll, J. 1990. "The Humean Tradition." *Philosophical Review* 99:185–219.

Cartwright, N. 1983. *How the Laws of Physics Lie*. Oxford.

Cesi, B. 1636. *Mineralogia, sive Naturalis philosophiae thesauri, in quibus metallicae concretionis medicatorúmque fossilium miracula, terrarum pretium, colorum & pigmentorum apparatus, concretorum succorum virtus, lapidum atque gemmarum dignitas continentur, &c*. Leiden.

Cavallo, G. et al. 1989–. *Lo spazio letterario di Roma antica*. Rome.

Chakravartty, A. 1998. "Semirealism." *Studies in the History and Philosophy of Science* 29:391–408.

———. 2007. *A Metaphysics for Scientific Realism: Knowing the Unobservable*. Cambridge.

Chang, H. 2004. *Inventing Temperature*. Oxford.

———. 2009. "We Have Never Been Whiggish (about Phlogiston)." *Centaurus* 51:239–64.

Cheyne, C., and J. Worrall, eds. 2006. *Rationality and Reality: Conversations with Alan Musgrave*. New York.

Chinn, C. A., and W. F. Brewer. 1993. "The Role of Anomalous Data in Knowledge Acquisition: A Theoretical Framework and Implications for Science Instruction." *Review of Educational Research* 63:1–49.

Churchland, P. M. 1979. *Scientific Realism and the Plasticity of Mind*. Cambridge.

———. 1988. "Perceptual Plasticity and Theoretical Neutrality." *Philosophy of Science* 55:167–87.

Coady, C. A. J. 1992. *Testimony*. Oxford.

Codoñer Marino, C., ed. 1979. *L. Annaei Senecae naturales quaestiones*. Madrid.

Cohen, S. M. 1978. "Essentialism in Aristotle." *Review of Metaphysics* 31:387–405.

Cohen, T. 1988. "What's Special about Photography?" *Monist* 71:292–305.

Collins, R. 1994. "Against the Epistemic Value of Prediction over Accommodation." *Nous* 28:210–24.

Connolly, J. 2007. *The State of Speech: Rhetoric and Political Thought in Ancient Rome*. Princeton.

Corbeill, A. 2001. "Education in the Roman Republic: Creating Traditions." In Too 2001, 261–88.

Corcoran, T. H. 1971. *Seneca: Natural Questions*. Loeb ed. Cambridge, MA.

Creager, A. N. H., E. Lunbeck, and M. N. Wise, eds. 2007. *Science without Laws: Model Systems, Cases, Exemplary Narratives*. Durham, NC.

Crombie, A. 1994. *Styles of Scientific Thinking in the European Tradition*. London.

Cuomo, S. 2001. *Ancient Mathematics*. London.
———. 2007. *Technology and Culture in Greek and Roman Antiquity*. Cambridge.
Currie, G. 1991. "Photography, Painting, and Perception." *Journal of Aesthetics and Art Criticism* 49:23–29.
Daston, L. 1988. *Classical Probability in the Enlightenment*. Princeton.
Daston, L., and M. Stolleis, eds. 2006. *Natural Law and Laws of Nature in Early Modern Europe*. Aldershot.
Daston, L., and F. Vidal, eds. 2004. *The Moral Authority of Nature*. Chicago.
Davidson, D. 1983. "A Coherence Theory of Truth and Knowledge." In D. Henrich, ed., *Kant oder Hegel?* Stuttgart, 423–38.
———. 2001. *Subjective, Intersubjective, Objective*. Oxford.
Davis, P. J. 2006. *Ovid and Augustus: A Political Reading of Ovid's Erotic Poems*. London.
Dear, P. 1995. *Discipline and Experience*. Chicago.
Delaney, D. 2003. *Law and Nature*. Cambridge.
della Porta, G. 1589. *Magiae naturalis libri XX. Ab ipso authore expurgati, & superaucti, in quibus scientiarum naturalium diuitiae, & delitiae demonstrantur*. Naples.
———. 1658. *Natural Magick: In XX Bookes by Iohn Baptist Porta, a Neapolitane*, Trans. T. Young and S. Speed. London.
Denyer, N. 1985. "The Case against Divination: An Examination of Cicero's *De divinatione*." *Proceedings of the Cambridge Philological Society*, n.s., 31:1–10.
———. 1994. "Why Do Mirrors Reverse Left/Right and Not Up/Down?" *Philosophy* 69:205–10.
Descartes, R. 1664. *Le monde de Mr Descartes, ou, le traité de la lumière, et des autres principaux objects des sens*. Paris.
Descola, P. 2004. "Constructing Natures: Symbolic Ecology and Social Practice." In P. Descola and G. Pálsson, eds., *Nature and Society: Anthropological Perspectives*. London, 82–102.
Detienne, Marcel. 1981. *L'invention de la mythologie*. Paris.
Devitt, M. 2006. "Worldmaking Made Hard." *Croatian Journal of Philosophy* 6:3–25.
Dillon, J. M., and A. A. Long, eds. *The Question of "Eclecticism."* Berkeley.
Domenicucci, P. 1996. *Astra Caesarum: Astrologia e catasterismo da Cesare a Domiziano*. Pisa.
Dominik, W. J. 1997. "The Style Is the Man." In W. J. Domink, ed., *Roman Eloquence*. London, 50–70.
Dominik, W. J., and J. Hall, eds. 2007. *A Companion to Roman Rhetoric*. Oxford.
Donini, P. 1988. "The History of the Concept of Eclecticism." In Dillon and Long, 15–33.
———. 1992. "Galeno e la filosofia." *Aufstieg und Niedergang der römischen Welt* II.36.5:3484–3505.
Doody, A. 2009. "Pliny's *Natural History*: *Enkuklios Paideia* and the Ancient Encyclopedia." *Journal of the History of Ideas* 70:1–21.
Douglas, A. E. 1995. "Form and Content in the *Tusculan Disputations*." In J. G. F. Powell, ed., *Cicero the Philosopher*. Oxford, 197–218.
Douglas, M. 1966. *Purity and Danger*. New York.
———. 1975. *Implicit Meanings*. London.
Dretske, F. 1977. "Laws of Nature." *Philosophy of Science* 44:248–68.

———. 1995. "Perception." In R. Audi, ed. *Cambridge Dictionary of Philosophy*. Cambridge, 657.

Duhem, P. 1906. *La théorie physique*. Paris.

Earman, J. 1984. "Laws of Nature: The Empiricist Challenge." In R. Bogdan, ed., *D. M. Armstrong*. Dordrecht.

Eastwood, B. S. 2002. *The Revival of Planetary Astronomy in Carolingian and Post-Carolingian Europe*. London.

Ellis, B. 1988. "Internal Realism." *Synthese* 76:409–34.

Erler, M., and M. Schofield. 1999. "Epicurean Ethics." In Algra et al. 1999, 642–74.

Evans, J. 2004. "The Astrologer's Apparatus: A Picture of Professional Practice in Greco-Roman Egypt." *Journal for the History of Astronomy* 35:1–44.

Evans, J., and J. L. Berggren. 2006. *Geminus: An Introduction to the Phenomena*. Princeton.

Evans, J., C. Carman, and A. Thorndike. 2010. "Solar Anomaly and Planetary Displays in the Antikythera Mechanism." *Journal for the History of Astronomy* 41:1–39.

Feeney, D. 1998. *Literature and Religion at Rome: Cultures, Contexts, and Beliefs*. Cambridge.

Falconer, W. A. 1923. *Cicero, de senectute, de amicitia, de divinatione*. Loeb ed. Cambridge.

Ferejohn, M. 1991. *The Origins of Aristotelian Science*. New Haven.

Fernandez, J. W. 1990. "Tolerance in a Repugnant World and Other Dilemmas in the Cultural Relativism of Melville J. Herskovits." *Ethos* 18:140–64.

Feyerabend, P. 1975. *Against Method: Outline of an Anarchistic Theory of Knowledge*. London.

Fitzgerald, W. 2000. "The Case for Rome." *Classical Philology* 95:207–19.

Flemming, R. 2000. *Medicine and the Making of Roman Women*. Oxford.

———. 2003. "Empires of Knowledge: Medicine and Health in the Hellenistic World." In A. Erskine, ed., *A Companion to the Hellenistic World*. Oxford, 449–63.

———. 2007a. "Women, Writing, and Medicine in the Classical World." *Classical Quarterly* 57:1–23.

———. 2007b. "Galen's Imperial Order of Knowledge." In König and Whitmarsh, 2007, 241–77.

Fodor, J. 1983. *The Modularity of Mind*. Cambridge, MA.

———. 1984. "Observation Reconsidered." *Philosophy of Science* 51:23–43.

———. 1988. "A Reply to Churchland's 'Perceptual Plasticity and Theoretical Neutrality.'" *Philosophy of Science* 55:188–98.

Formisano, M. 2001. *Tecnica e scrittura*. Rome.

———. 2003. "Auctor, utilitas, princeps." *Voces* 14:155–64.

———. 2004. "Aspetti letterari della cultura della tecnica tardoromana." In A. Marcone, ed., *Società e cultura in età tardoantica*. Florence, 77–90.

Foster, J. 2000. *The Nature of Perception*. Oxford.

Foucault, M. 1966. *Les mots et les choses: Un archéologie des sciences humaines*. Paris.

Fowler, D. 1989, "*Lucretius* and Politics." In Griffin and Barnes 1989, 120–50.

Fox, M. 2007. *Cicero's Philosophy of History*. Oxford.

Franklin, J. 2001. *The Science of Conjecture*. Baltimore.

Frede, M. 1981. "On Galen's Epistemology." In Nutton 1981, 65–86.

———. 1999. "Stoic Epistemology." In Algra et al. 1999, 295–322.

Freeth, T. et al. 2006. "Decoding the Ancient Greek Astronomical Calculator Known as the Antikythera Mechanism." *Nature* 444 (30 November): 587–91.

French, R. 1994. *Ancient Natural History*. London.

French, R., and F. Greenaway, eds. 1986. *Science in the Early Roman Empire*. London.

Friedman, M. 1974. "Explanation and Scientific Understanding." *Journal of Philosophy* 71:5–19.

Friedman, C., and U. L. Abbas. 2003. "Is Medical Informatics a Mature Science?" *International Journal of Medical Informatics* 69:261.

Gagarin, M. 2002. *Antiphon the Athenian: Oratory, Law, and Justice in the Age of the Sophists*. Austin, TX.

Gale, M. R. 2000. *Virgil on the Nature of Things*. Cambridge.

Galison, P. 1987. *How Experiments End*. Chicago.

———. 1997. *Image and Logic*. Chicago.

Gallagher, R. L. 2001. "Metaphor in Cicero's 'De Re Publica.'" *Classical Quarterly* 51:509–19.

Getz, W. M. 2003. "Is Population Ecology a Mature Science?" *BioScience* 53:885–88.

Gelman, S. A., and E. M. Markman. 1986. "Categories and Induction in Young Children." *Cognition* 23:183–209.

Gibson, J. J. 1966. *The Senses Considered as Perceptual Systems*. Boston.

Giere, R. N. 1983. "Testing Theoretical Hypotheses." *Minnesota Studies in the Philosophy of Science* 10:269–98.

———. 1999. *Science without Laws*. Chicago.

Gilbert, W. 1600. *De magnete, magneticisque corporibus, et de magno magnete tellure; Physiologica noua, plurimis & argumentis, & experimentis demonstrata*. London.

Gill, C. 2003. "The School in the Roman Imperial Period." In B. Inwood, *The Cambridge Companion to the Stoics*. Cambridge, 33–58.

Gleason, M. 1995. *Making Men: Sophists and Self-Presentation in Ancient Rome*. Princeton.

———. 2009. "Shock and Awe: The Performance Dimension of Galen's Anatomy Demonstrations." In C. Gill, T. Whitmarsh, and J. Wilkins, eds., *Galen and the World of Knowledge*. Cambridge, 85-114.

Goar, R. J. 1968. "The Purpose of the De divinatione." *Transactions and Proceedings of the American Philological Association* 99:241–48.

Goldberg, S. M. 1997. "Melpomene's Declamation." In W. Dominik, ed., *Roman Eloquence*. London, 166–81.

Goldhill, S. 2001a. "Introduction: Setting an Agenda." In S. Goldhill, ed., *Being Greek under Rome*. Cambridge, 1–28.

———. 2001b. "The Erotic Eye." In S. Goldhill, ed., *Being Greek under Rome*. Cambridge, 154–94.

———. 2008. *The End of Dialogue in Antiquity*. Cambridge.

Goldstein, B. R. 1967. "The Arabic Version of Ptolemy's *Planetary Hypotheses*." *Transactions of the American Philosophical Society* 57:4–12.

Goodman, N. 1955. *Fact, Fiction, and Forecast*. Cambridge, MA.

———. 1968. *Languages of Art*. Indianapolis.

———. 1972. "Seven Strictures on Similarity." In Goodman, *Problems and Projects*. Indianapolis, 437–46.
———. 1978. *Ways of Worldmaking*. 4th ed. Indianapolis.
Gordin, M. D. 2004. *A Well-Ordered Thing: Dmitrii Mendeleev and the Shadow of the Periodic Table*. New York.
Grafton, A. T., and N. Swerdlow. 1988. "Calendar Dates and Ominous Days in Ancient Historiography." *Journal of the Warburg and Courtauld Institutes* 51:14–42.
Graham, D. 2006. *Explaining the Cosmos*. Princeton.
Griffin, M., and J. Barnes, eds. 1989. *Philosophia Togata*. Oxford.
Guillaumont, F. 1984. *Philosophe et augure: Recherches sur la théorie cicéronienne de la divination*. Brussels.
Gundel, H. G. 1975. "Claudius." In K. Ziegeler and W. Sontheimer, eds., *Der Kleine Pauly*. Munich.
Gunderson, E. 2003. *Declamation, Paternity, and Roman Identity*. Cambridge.
Guthrie, W. K. C. 1962–81. *A History of Greek Philosophy*. Cambridge.
Haack, S. 1993. *Evidence and Inquiry*. Oxford.
Habinek, T. 1998. *The Politics of Latin Literature*. Princeton.
Habinek, T., and A. Schiesaro, eds. 1997. *The Roman Cultural Revolution*. Cambridge.
Hacking, I. 1975. *The Emergence of Probability*. Cambridge.
———. 1979. "Michel Foucault's Immature Science." *Nous* 13:39–51.
———. 1982. "Language, Truth, and Reason." In Hollis and Lukes 1982, 48–66.
———. 1983. *Representing and Intervening*. Cambridge.
———. 1990. *The Taming of Chance*. Cambridge.
———. 1993. "Working in a New World: The Taxonomic Solution." In Horwich 1993, 275–310.
———. 1999. *The Social Construction of What?* Cambridge, MA.
———. 2002. *Historical Ontology*. Cambridge, MA.
Hankinson, R. J. 1989. "Galen and the Best of All Possible Worlds." *Classical Quarterly* 39:206–27.
———. 1991. "Galen's Anatomy of the Soul." *Phronesis* 36:197–233.
———. 1992. "Galen's Philosophical Eclecticism." *Aufstieg und Niedergang der römischen Welt* II.36.5:3505–22.
———. 2008. "Epistemology." In R. J. Hankinson, ed., *The Cambridge Companion to Galen*. Cambridge, 157–83.
Hanson, A. E. 1987. "The Eight-Months' Child and the Etiquette of Birth: *Obsit Omen!*" *Bulletin of the History of Medicine* 61:589–602.
Hanson, N. R. 1958. *Patterns of Discovery*. Cambridge.
Hardie, P. 1986. *Virgil's* Aeneid: *Cosmos and Imperium*. Oxford.
Heinimann, F. 1945. *Nomos und Physis*. Basel.
Hempel, C. 1965. *Aspects of Scientific Explanation*. New York.
Hempel, C., and P. Oppenheim. 1948. "Studies in the Logic of Explanation." *Philosophy of Science* 15:135–75.
Herskovits, M. J. 1972. *Cultural Relativism: Perspectives in Cultural Pluralism*. New York.
Hine, H., ed. 1996. *L. Annaei Senecae Naturalium quaestionum libros*. Leipzig.
Hollis, M., and S. Lukes, eds. 1982. *Rationality and Relativism*. Oxford.

Horster, M., and C. Reitz, eds. 2003. *Antike Fachschriftsteller*. Stuttgart.
Horstmanshoff, M. 2004. "Aelius Aristides: A Suitable Case for Treatment." In Borg and Borg 2004, 277–90.
Horwich, P., ed. 1993. *World Changes*. Cambridge, MA.
Howson, C., and A. Franklin. 1991. "Maher, Mendeleev, and Baysianism." *Philosophy of Science* 58:574–85.
Huby, P., and G. Neal, eds. 1989. *The Criterion of Truth*. Liverpool.
Hunger, H., and D. Pingree. 1999. *Astral Sciences in Mesopotamia*. Leiden.
Ierodiakonou, K. 2002. "Aristotle's Use of Examples in the *Prior Analytics*." *Phronesis* 47:127–52.
Iliffe, R. 2007. *Newton: A Very Short Introduction*. Oxford.
Ilting, K. H. 1983. *Naturrecht und Sittlichkeit*. Stuttgart.
Inwood, B. 1987. "Commentary on Striker." *Proceedings of the Boston Area Colloquium in Ancient Philosophy* 2:95–101.
———. 2005. *Reading Seneca*. Oxford.
Inwood, B., and P. Donini. 1999. "Stoic Ethics." In Algra et al. 1999, 675–738.
Jäger, C. 2007. "Is Coherentism Coherent?" *Analysis* 67:341–44.
Jaeger, F. 1910. *De oraculis quid veteres philosophi iudicaverint*. Bonn.
James, W. 1907. *Pragmatism*. New York. Repr. 1975 in F. H. Burkhardt, ed., *The Works of William James*, vol. 1. Cambridge, MA.
Jeyes, U. 1989. *Old Babylonian Extispicy*. Istanbul.
Johansen, T. K. 2004. *Plato's Natural Philosophy*. Cambridge.
Johnston, S. I., and P. Struck, eds. 2005. *Mantikê: Studies in Ancient Divination*. Leiden.
Johnsson, E. M. 1995. *Le miroir: Naissance d'un genre littéraire*. Paris.
Jones, A. 1999. *Astronomical Papyri from Oxyrhynchus*. Philadelphia.
———. 2005. "Ptolemy's Mathematical Models and Their Meaning." In G. Van Brummelen and M. Kinyon, eds., *Mathematics and the Historian's Craft: The Kenneth O. May Lectures*. New York.
Jouanna, J. 2003. "La notion de nature chez Galien." In Barnes and Jouanna 2003, 229–68.
Kahn, C. H. 2001. *Pythagoras and the Pythagoreans: A Brief History*. Indianapolis.
Kany-Turpin, J. 2004. *Cicéron: De la divination, De diuinatione*. Paris.
Karenberg, A., and F. P. Moog. 2003. "Next Emperor, Please! No End to Retrospective Diagnostics." *Journal of the History of the Neurosciences* 13:143–49.
Kaster, R. A. 2001. "Controlling Reason: Declamation in Rhetorical Education at Rome." In Too 2001, 317–38.
Kelley, D. R. 1990. *The Human Measure*. Cambridge, MA.
Kennedy, G. 1972. *The Art of Rhetoric in the Roman World*. Princeton.
Kerferd, G. B. 1981. *The Sophistic Movement*. Cambridge.
King, H. 1998. *Hippocrates' Woman: Reading the Female Body in Ancient Greece*. London.
Kitcher, P. 1989. "Explanatory Unification and the Causal Structure of the World." In Kitcher and Salmon, 410–505.
———. 1993. *The Advancement of Science*. Oxford.
———. 2001. "Real Realism: The Galilean Strategy." *Philosophical Review* 110:151–97.
Kitcher, P., and W. Salmon, eds. 1989. *Scientific Explanation* (Minnesota Studies in the Philosophy of Science 13). Minneapolis.

Koch-Westenholz, U. 1995. *Mesopotamian Astrology*. Copenhagen.

———. 2000. *Babylonian Liver Omens*. Copenhagen.

Kollesch, J. 1981. "Galen und die Zweite Sophistik." In Nutton 1981, 1–12.

König, J., and T. Whitmarsh, eds. 2007. *Ordering Knowledge in the Roman Empire*. Cambridge.

Krajewski, W. 1977. *Correspondence Principle and Growth of Science*. Dordrecht.

Krostenko, B. A. 2000. "Beyond (Dis)belief: Rhetorical Form and Religious Symbol in Cicero's *De divinatione*." *Transactions of the American Philological Association* 130:353–91.

Krüger, L. 1981. "Vergängliche Erkenntnis der beharrenden Natur." In H. Poser, ed., *Wandel des Vernunftbegriffs*. Munich, 223–49.

———. 1991. "Does Progress Lead to Scientific Truth?" In E. Deutsch, ed., *Culture and Modernity*. Honolulu, 615–28.

Krüger, L., L. J. Daston, and M. Heidelberger, eds. 1987. *The Probabilistic Revolution*. Cambridge, MA.

———. 1981. "Galen's Religious Belief." In Nutton 1981, 117–30.

Kuhn, T. S. 1962. "The Structure of Scientific Revolutions." *International Encyclopedia of Unified Science: Foundations of the Unity of Science*, vol. 2, no. 2. Chicago.

———. 1970. *The Structure of Scientific Revolutions*. 2nd ed. Chicago.

———. 1987. "What Are Scientific Revolutions?" In Kruger, Daston, and Heidelberger 1987, 7–22.

———. 1993. "Afterwords." In Horwich 1993, 311–41.

Kuriyama, S. 2002. *The Expressiveness of the Body*. New York.

Kuukkanen, J.-M. 2007. "Kuhn, the Correspondence Theory of Truth and Coherentist Epistemology." *Studies in the History and Philosophy of Science* 38:555–66.

Kusch, M. 2002. *Knowledge by Agreement*. Oxford.

Kusch, M., and P. Lipton. 2002. "Testimony, a Primer." *Studies in History and Philosophy of Science* 33:209–17.

Lakatos, I. 1970. "Falsification and the Methodology of Scientific Research Programmes." In I. Lakatos and A. Musgrave, eds., *Criticism and the Growth of Knowledge*. Cambridge, 91–196.

Lange, M. 2000. *Natural Laws in Scientific Practice*. Oxford.

Langholf, V. 1990. *Medical Theories in Hippocrates: Early Texts and the "Epidemics."* Berlin.

Larsen, M. T. 1987. "The Mesopotamian Lukewarm Mind: Reflections on Science, Divination and Literacy." In F. Rochberg-Halton, ed., *Language, Literature, and History: Philological and Historical Studies Presented to Erica Reiner*. New Haven, 203–25.

Latour, B. 1991. *We Have Never Been Modern*. C. Porter, trans. Cambridge, MA.

Laudan, L. 1981a. "A Confutation of Convergent Realism." *Philosophy of Science* 48:19–49.

———. 1981b. "The Pseudo-Science of Science." *Philosophy of the Social Sciences* 11:173–98.

Lehoux, D. 1999. "All Voids Large and Small, Being a Discussion of Place and Void in Strato of Lampsacus's Matter Theory." *Apeiron* 32:1–36.

———. 2003a. "Tropes, Facts, and Empiricism." *Perspectives on Science* 11:326–45.

———. 2003b. "The Historicity Question in Mesopotamian Divination." In J. Steele and A. Imhausen, eds., *Under One Sky: Astronomy and Mathematics in the Ancient Near East*. Munster.

———. 2006a. "Laws of Nature and Natural Laws." *Studies in History and Philosophy of Science* 37:527–49.

———. 2006b. "Tomorrow's News Today: Astrology, Fate, and the Ways Out." *Representations* 95:105–22.

———. 2007a. *Astronomy, Weather, and Calendars in the Ancient World*. Cambridge.

———. 2007b. "Observers, Objects, and the Embedded Eye; Or, Seeing and Knowing in Ptolemy and Galen." *ISIS* 98:447–67.

Lewis, D. 1983. "New Work for a Theory of Universals." *Australasian Journal of Philosophy* 61:343–77.

———. 1996. "Elusive Knowledge." *Australasian Journal of Philosophy* 74:549–67.

———. 2001. "Forget about 'The Correspondence Theory of Truth.'" *Analysis* 61:275–80.

Li Causi, P. 2003. *Sulle tracce del manticora*. Palermo.

———. 2008. *Generare in comune*. Palermo.

Lindberg, D. 1976. *Theories of Vision from al-Kindi to Kepler*. Chicago.

———. 1992. *The Beginnings of Western Science*. Chicago.

Linderski, J. 1982. "The Augural Law." *Aufstieg und Niedergang der römischen Welt* II.16.3:2146–2312.

———. 1986. "Watching the Birds: Cicero the Augur and the Augural *Templa*." *Classical Philology* 81:330–40.

Lipton, P. 1998. "The Epistemology of Testimony." *Studies in History and Philosophy of Science* 29:1–31.

Lloyd, G. E. R. 1966. *Polarity and Analogy*. Cambridge.

———. 1979. *Magic, Reason, and Experience*. Cambridge.

———. 1987. *Revolutions of Wisdom*. Berkeley.

———. 1990. *Demystifying Mentalities*. Cambridge.

———. 2002. *The Ambitions of Curiosity*. Cambridge.

———. 2004. *Ancient Worlds, Modern Reflections*. Oxford.

———. 2007. *Cognitive Variations*. Oxford.

Loewer, B. 1996. "Humean Supervenience." *Philosophical Topics* 24:101–26.

Long, A. A. 1982. "Astrology: Arguments Pro and Contra." In J. Barnes et al., eds., *Science and Speculation*. Cambridge, 165–92.

———. 1986. "Pleasure and Social Utility—The Virtues of Being Epicurean." In H. Flashar and O. Gigon, eds., *Aspects de la philosophie hellénistique*. Geneva, 283–324.

———. 1989. "Ptolemy on the Criterion: An Epistemology for the Practicing Scientist." In Huby and Neal 1989, 151–78.

———. 1999. "Stoic Psychology." In Algra et al. 1999, 560–84.

———. 2003. "Roman Philosophy." In D. Sedley, ed., *The Cambridge Companion to Greek and Roman Philosophy*. Cambridge, 184–210.

Long, L. 2001. *Openness, Secrecy, Authorship*. Baltimore.

Lyons, T. D. 2002. "Scientific Realism and the Pessimistic Meta-Modus Tollens." In S. Clarke and T. D. Lyons, eds., *Recent Themes in Philosophy of Science*. Dordrecht, 63–90.

———. 2006. "Scientific Realism and the Stratagema de Divide et Impera." *British Journal for the Philosophy of Science* 57:537–60.

MacKendrick, P. L. 1989. *The Philosophical Books of Cicero.* London.

Madsen, J. M. 2009. *Eager to Be Roman.* London.

Maher, P. 1988. "Prediction, Accommodation, and the Logic of Discovery." *Proceedings of the Philosophy of Science Association* 1:273–85.

———. 1990. "How Prediction Enhances Confirmation." In M. Dunn and A. Gupta, eds., *Truth or Consequences.* Dordrecht.

Manuli, P. 1981. "Claudio Tolomeo: Il criterio e il principio." *Rivista critica di storia della filosofia* 35:64–88.

Marcus, R. 1971. "Essential Attribution." *Journal of Philosophy* 68:187–202.

Mattern, S. P. 1999. "Physicians and the Roman Imperial Aristocracy: The Patronage of Therapeutics." *Bulletin of the History of Medicine* 73:1–18.

———. 2008. *Galen and the Rhetoric of Healing.* Baltimore.

Matthews, G. 1990. "Aristotelian Essentialism." *Philosophy and Phenomenological Research* 1, suppl., 251–62.

Matthiolus, P. 1554. *Petri Andreae Matthioli medici senensis Commentarii, in libros sex Pedacii Dioscoridis Anazarbei, De medica materia. Adiectis quam plurimis plantarum & animalium imaginibus, eodem authore.* Venice.

Maul, S. 1999. "How the Babylonians Protected Themselves against Calamities Announced by Omens." In T. Abusch and K. van der Toorn, eds., *Mesopotamian Magic.* Groningen, 123–30.

Maurach, G. 1989. *Geschichte der römischen Philosophie.* Darmstadt.

May, M. T., trans. and comm. 1968. *Galen, on the Usefulness of the Parts of the Body.* Ithaca, NY.

Mayo, D. 1996. *Error and the Growth of Experimental Knowledge.* Chicago.

McCluskey, S. 1998. *Astronomies and Cultures in Early Medieval Europe.* Cambridge.

McKirahan, R. 1992. *Principles and Proofs: Aristotle's Theory of Demonstrative Science.* Princeton.

Medin, D. L. 1989. "Concepts and Conceptual Structure." *American Psychologist* 44:1469–81.

Meissner, B. 1999. *Die technologische Fachliteratur der Antike.* Berlin.

Menghi, M., and M. Vegetti. 1984. *Le passioni e gli errori dell' anima.* Venice.

Miller, F. D. 1991. "Aristotle on Natural Law and Justice." In D. Keyt and F. D. Miller, eds., *A Companion to Aristotle's Politics.* Oxford, 279–306.

Milton, J. R. 1981. "The Origin and Development of the Concept of the 'Laws of Nature.'" *Archives Européennes de Socicologie* 22:173–95.

———. 1998. "Laws of Nature." In D. Garber and M. Ayers, eds., *The Cambridge History of Seventeenth-Century Philosophy.* Cambridge, 680–701.

Mitchell, S. 2000. "Dimensions of Scientific Law." *Philosophy of Science* 67:242–65.

Moatti, C. 1997. *La raison de Rome.* Paris.

Morford, M. 2002. *The Roman Philosophers.* London.

Morgan, T. 1998. *Literate Education in the Hellenistic and Roman Worlds.* Cambridge.

Mumford, S. 2004. *Laws in Nature.* New York.

Murphy, G., and D. Medin. 1985. "The Role of Theories in Conceptual Coherence." *Psychological Review* 92:289–316.

Murphy, T. 1998. "Cicero's First Readers: Epistolatory Evidence for the Dissemination of His Works." *Classical Quarterly*, n.s., 48:492–505.

Musgrave, A. 1974. "Logical versus Historical Theories of Confirmation." *British Journal for the Philosophy of Science* 25:1–23.

———. 1981. "Der Mythos vom Instramentalismus in der Astronomie." In H. P. Duerr, ed., *Versuchungen: Aufsätze zur Philosophie Paul Feyerabends*, vol. 2. Frankfurt, 231–79.

———. 1988. "The Ultimate Argument for Scientific Realism." In R. Nola, ed., *Relativism and Realism in Science*. Boston, 229–52.

———. 1992. "Realism about What?" *Philosophy of Science* 59:691–97.

Neri, V. 1982. "Dei, Fato e divinazione nella letteratura latina del I sec. d. C." *Aufstieg und Niedergang der römischen Welt* II.16.3:1974–2051.

Neugebauer, O. 1951. "The Study of Wretched Subjects." *Isis* 42:111.

———. 1975. *A History of Ancient Mathematical Astronomy*. New York.

Neugebauer, O., and H. B. van Hoesen. 1959. *Greek Horoscopes*. Philadelphia.

Neurath, O. 1932–33. "Protokollsätze." *Erkenntnis* 3:204–14.

Nicolet, C., ed. 1996. *Les littératures techniques dans l'antiquité romaine*. Geneva.

North, J. 1986. "Religion and Politics, from Republic to Principate." *Journal of Roman Studies* 76:251–58.

———. 1990. "Diviners and Divination at Rome." In M. Beard and J. North, eds., *Pagan Priests: Religion and Power in the Ancient World*. Ithaca, 51–71.

Nussbaum, M. 1994. *The Therapy of Desire*. Princeton.

Nutton, V., ed. 1981. *Galen: Problems and Prospects*. London.

———. 2004. *Ancient Medicine*. New York.

Olson, M. 1983. "Towards a Mature Social Science." *International Studies Quarterly* 27:29–37.

Olsson, E. J., and T. Shogenji. 2004. "Can We Trust Our Memories? C. I. Lewis's Coherence Argument." *Synthese* 142:21–41.

Oppenheim, A. L. 1977. *Ancient Mesopotamia*. Rev. ed. Chicago.

Orr, D. G. 1978. "Roman Domestic Religion: The Evidence of the Household Shrines." *Aufstieg und Niedergang der römischen Welt* II.16.2:1557–91.

Ostwald, M. 1986. *From Popular Sovereignty to the Sovereignty of Law: Law, Society, and Politics in Fifth-Century Athens*. Berkeley.

Paniagua Aguilar, D. 2006. *El panorama literario técnico-científico en Roma*. Salamanca.

Park, K., and L. Daston. 2006. "Introduction." In K. Park and L. Daston, eds., *The Cambridge History of Science*, vol. 3, *Early Modern Science*. Cambridge, 1–21.

Parker, G. 2008. *The Making of Roman India*. Cambridge.

Peachkin, M. 2004. *Frontinus and the* Curae *of the* Curator Aquarum. Stuttgart.

Pease, A. S. 1979 [1920]. *M. Tulli Ciceronis de divinatione*. New York.

Pigeaud, J. 1993. "Les problèmes de la création chez Galien." In J. Kollesch and D. Nickel, eds., *Galen und das hellenistische Erbe*. Stuttgart.

Pigeaud, A., and J. Pigeaud, eds. 2000. *Les textes médicaux latins comme littérature*, Nantes.

Plat, H. 1653. *The Jevvel House of Art and Nature: Containing Divers Rare and Profitable Inventions, together with sundry new Experiments in the Art of Husbandry . . . Whereunto is added, A rare and excellent Discourse of Minerals, Stones, Gums, and Rosins; with the vertues and use thereof, By D. B. Gent.* London.

Pocock, J. G. A. 1985. *Virtue, Commerce, and History.* Cambridge.

Pollan, M. 2001. *The Botany of Desire.* New York.

Potter, D. 1994. *Prophets and Emperors.* Cambridge, MA.

Powell, J. G. F. 1990. *Laelius: On Friendship and the Dream of Scipio.* Oxford.

———. 1995a. "Cicero's Translations from Greek." In J. G. F. Powell, ed., 273–300.

———, ed. 1995b. *Cicero the Philosopher.* Oxford.

———. 1996. "Second Thoughts on the Dream of Scipio." *Papers of the Leeds International Latin Seminar* 9:13–27.

———. 2006. *M. Tulli Ciceronis: De republica, de legibus, Cato maior de senectute, Laelius de amicitia.* Oxford.

Preyer, G., and G. Peter, eds. 2005. *Contextualism in Philosophy.* Oxford.

Proctor, R. N. 2007. "'-Logos,' '-ismos,' and '-ikos': The Political Iconicity of Denominative Suffixes in Science (or, Phonesthemic Tints and Taints in the Coining of Science Domain Names)." *Isis* 98:290–309.

Psillos, S. 1999. *Scientific Realism.* New York.

———. 2005. "Scientific Realism and Metaphysics." *Ratio* 18:385–404.

Putnam, H. 1962. "What Theories Are Not." In E. Nagel, A. Tarski and P. Suppes, eds., *Logic, Methodology, and Philosophy of Science.* Stanford, 240–51.

———. 1975. *Mathematics, Matter, and Method.* Cambridge.

———. 1978. *Meaning and the Moral Sciences.* London.

———. 1999. *The Threefold Cord.* New York.

Quine, W. V. 1953. "Two Dogmas of Empiricism." In *From a Logical Point of View.* Cambridge, MA.

Quine, W. V., and J. S. Ullian. 1970. *The Web of Belief.* New York.

Rabelais, F. 1653. *The Works of Francis Rabelais, Doctor in Physick: Containing Five Books of the Lives, Heroick Deeds, and Sayings of Gargantua, and His Sonne Pantagruel, Together with the Pantagrueline Prognostication, the Oracle of the Divine Bacbuc, and Response of the Bottle.* Trans. T. Urquhart and P. A. Motteux. London.

Ramsay, J. T., and A. L. Licht. 1997. *The Comet of 44 BC and Caesar's Funeral Games.* Atlanta.

Rasmussen, S. W., 1998. "Cicero's Stand on Prodigies: A Non-Existent Dilemma?" In R. L. Wildfang and J. Isager, eds., *Divination and Portents in the Roman World.* Odense, 9–24.

Rehak, P. 2006. *Imperium and Cosmos.* Madison, WI.

———. 1995. "Astral Magic in Babylonia." *Transactions of the American Philosophical Society* 84:4.

Reiner, E., and D. Pingree. 1975–1998. *Babylonian Planetary Omens.* Malibu and Groningen.

Revell, L. 2009. *Roman Imperialism and Local Identities.* Cambridge.

Risselada, R. 1993. *Imperatives and Other Directive Expressions in Latin.* Amsterdam.

Ritter, J. 1995. "Babylon—1800." In M. Serres, ed., *A History of Scientific Thought*. London.

———. 2005. "Reading Strasbourg 368: A Thrice-Told Tale." *Boston Studies in the Philosophy of Science* 238:177–200.

Rives, J. 2007. *Religion in the Roman Empire*. Oxford.

Roberts, J. 1999. "'Laws of Nature' as an Indexical Term: A Reinterpretation of Lewis's Best-System Analysis." *Philosophy of Science*, 66:S502–11.

Rocca, J. 2003. *Galen on the Brain*. Leiden.

Rochberg, F. 2004. *The Heavenly Writing*. Cambridge.

Rohault, J. 1671. *Traité de physique*. Paris.

———. 1735. *Rohault's System of Natural Philosophy, Illustrated with D[r]. Samuel Clarke's Notes Taken mostly out of Sir Isaac Newton's Philosophy, Done into English By John Clarke, D. D. Dean of Sarum, the Third Edition*. London.

Rogers, A. J. 1960. "Mature Science—Retarded Profession." *Florida Entomologist* 43:155–62.

Roller, M. 2001. *Constructing Autocracy*. Princeton.

Rorty, R. 1991. "Pragmatism, Davidson, and Truth." *Philosophical Papers*, vol. 2. Cambridge, 126–50.

Routh, H. B., and K. R. Bhowmik. 1993. "History of Elephantiasis." *International Journal of Dermatology* 32:913–16.

Rowbottom, D. P. 2007. "A Refutation of Foundationalism?" *Analysis* 67:345–46.

Rowlands, M., M. T. Larsen, and K. Kristiansen, eds. 1987. *Centre and Periphery in the Ancient World*. Cambridge.

Ruby, J. E. 1986. "The Origins of Scientific 'Law.'" *Journal for the History of Ideas* 47, no. 3, 341–59.

Rüpke, J. 2001. *Die Religion der Römer*. Munich.

———., ed. 2007. *A Companion to Roman Religion*. Oxford.

Russell, B. 1921. *The Analysis of Mind*. London.

Russo, L. 1997. *La rivoluzione dimenticata*. Milan. Translated by S. Levy as *The Forgotten Revolution* (New York, 2000).

Sabra, A. I. 1989. *The Optics of Ibn al-Haytham*. London.

Saller, R. P. 1982. *Personal Patronage under the Early Empire*. Cambridge.

Sanocki, T. 1996. "Visions of a Mature Cognitive Science?" *American Journal of Psychology* 109:157–61.

Santini, C. et al., eds. 2002. *Letteratura scientifica e tecnica di Grecia e Roma*. Rome.

Schiavone, A. 1996. *La storia spezzata*. Rome. Translated by M. J. Schneider as *The End of the Past* (Cambridge, MA, 2000).

———. 2000. Trans. M. J. Schneider. *The End of the Past*. Cambridge, MA.

Schofield, M. 1986. "Cicero For and Against Divination." *Journal of Roman Studies* 76:47–64.

———. 1991. *The Stoic Idea of the City*. Cambridge.

———. 1999. "Social and Political Thought." In Algra, et al. 1999, 739–70.

———. 2008. "Ciceronian Dialogue." In Goldhill 2008, 63–84.

Scruton, R. 1981. "Photography and Representation." *Critical Inquiry* 7:577–603.

Sedley, D. 1989. "Philosophical Allegiance in the Greco-Roman World." In Griffin and Barnes 1989, 97–119.

———. 1998. *Lucretius and the Transformation of Greek Wisdom*. Cambridge.

———. 2007. *Creationism and Its Critics in Antiquity*. Berkeley.

Shapin, S. 1994. *A Social History of Truth*. Chicago.

———. 1996. *The Scientific Revolution*. Chicago.

Shogenji, T. 1999. "Is Coherence Truth Conducive?" *Analysis* 59:338–45.

Siebel, M., and W. Wolff. 2008. "Equivalent Testimonies as a Touchstone of Coherence Measures." *Synthese* 161:167–82.

Simon, G. 1988. *Le regard, l'être, et l'apparence dans l'optique de l'antiquité*. Paris.

Singer, P. N. 1997. "Levels of Explanation in Galen." *Classical Quarterly* 47:525–42.

Smith, A. M. 1988. "The Psychology of Visual Perception in Ptolemy's Optics." *ISIS* 79:188–207.

———. 1996. "Ptolemy's Theory of Visual Perception: An English Translation of the Optics." *Transactions of the American Philosophical Society* 86:2.

———. 1998. "The Physiological and Psychological Grounds of Ptolemy's Visual Theory: Some Methodological Considerations." *Journal of the Behavioral Sciences* 34:231–46.

———. 1999. "Ptolemy and the Foundations of Ancient Mathematical Optics: A Source-Based Guided Study." *Transactions of the American Philosophical Society* 89:3.

———. 2001. "Alhacen's Theory of Visual Perception." *Transactions of the American Philosophical Society* 91:4–5.

Snobelen, S. D. 2001. "'God of Gods, and Lord of Lords': The Theology of Isaac Newton's General Scholium to the *Principia*." *Osiris* 16:169–208.

Snyder, J., and N. Walsh Allen. 1975. "Photography, Vision, and Representation." *Critical Inquiry* 2:143–69.

Sober, E. 1997. "Two Outbreaks of Lawlessness in Recent Philosophy of Biology." *Philosophy of Science* 64:S458–67.

Sollberger, E. 1954–56. "Sur la chronologie des rois d'Ur et quelques problemes connexes." *Archiv für Orientforschung* 17.

Solomon, J. 2000. *Ptolemy's "Harmonics": Translation and Commentary.* Leiden.

Sperber, D. 1982. "Apparently Irrational Beliefs." In Hollis and Lukes 1982, 149–80.

Stahl, W. 1962. *Roman Science*. Madison, WI.

Stanford, P. K. 2006. *Exceeding our Grasp*. Oxford.

Stanley, J. 2005. *Knowledge and Practical Interests*. Oxford.

Starr, I. 1983. *The Rituals of the Diviner*. Malibu.

Strawson, P. F. 1979. "Perception and Its Objects." In G. Macdonald, ed., *Perception and Identity: Essays Presented to A. J. Ayer with His Replies*. Ithaca, NY, 41–60.

———. 1997. "Meaning and Context." *Entity and Identity and Other Essays*. Oxford, 216–31.

Stroup, S. C. 2007. "Greek Rhetoric Meets Rome: Expansion, Resistance, and Acculturation." In Dominik and Hall 2007, 23–37.

Stenholm, S. 1997. "Coming of Age in Quantum Optics." *Philosophical Transactions of the Royal Society of London* A, 355:2413–16.

Striker, G. 1987. "Origins of the Concept of Natural Law." *Proceedings of the Boston Area Colloquium in Ancient Philosophy* II:79–94.

Stoneham, T. 2007. "Is Coherentism Coherent?" *Analysis* 67:254–57.
Swain, S. 1996. *Hellenism and Empire: Language, Classicism, and Power in the Greek World AD 50–250*. Oxford.
Swerdlow, N., ed. 1999. *Ancient Astronomy and Celestial Divination*. Cambridge, MA.
Szemler, G. J. 1982. "Priesthoods and Priestly Careers." *Aufstieg und Niedergang der römischen Welt* II.16.3:2314–2331.
Takács, S. A. 1995. *Isis and Serapis in the Roman World*. Leiden.
Tambiah, S. J. 1990. *Magic, Science, Religion, and the Scope of Rationality*. Cambridge.
Tester, J. 1987. *A History of Western Astrology*. Woodbridge, UK.
Taub, L. 2008. *Aetna and the Moon*. Corvallis, OR.
Thagard, P. 1992. *Conceptual Revolutions*. Princeton.
———. 2000. *Coherence in Thought and Action*. Cambridge, MA.
———. 2006. "Testimony, Credibility, and Explanatory Coherence." *Erkenntnis* 63:295–316.
Thomas, R. 2000. *Herodotus in Context*. Cambridge.
Thorndike, L. 1923–1958. *A History of Magic and Experimental Science*. New York.
Tieleman, T. 1992. *Galen and Chrysippus: Argument and Refutation in the "De placitis" Books II–III*. Utrecht.
———. 2003. "Galen's Psychology." In Barnes and Jouanna 2003, 131–69.
Tierney, R. 2004. "The Scope of Aristotle's Essentialism in the Posterior Analytics." *Journal of the History of Philosophy* 42:1–20.
Too, Y. L. ed. 2001. *Education in Greek and Roman Antiquity*. Leiden.
Toohey, P. 2004. *Melancholy, Love, and Time,* Ann Arbor.
Toomer, G. J. 1984. *Ptolemy's Almagest*. London.
Tuchańska, B. 1992. "What Is Explained in Science?" *Philosophy of Science* 59:102–19.
Tybjerg, K. 2005. "Hero of Alexandria's Mechanical Treatises: Between Theory and Practice." In A. Schürmann, ed., *Physik-Mechanik: Geschichte der Mathematik und der Naturwissenschaften in der Antike*. Stuttgart, 204–26.
van Fraassen, B. C. 1980. *The Scientific Image*. Oxford.
———. 1989. *Laws and Symmetry*. Oxford.
———. 2002. *The Empirical Stance*. New Haven.
Van Nuffelen, P. 2010. "Varro's *Divine Antiquities:* Roman Religion as an Image of Truth." *Classical Philology* 105:162–88.
Verboven, K. 2002. *The Economy of Friends: Economic Aspects of Amicitia and Patronage in the Late Republic*. Brussels.
Veyne, P. 1983. *Les Grecs ont-ils cru à leurs mythes?* Paris.
Vicentini, A. 1634. *Alexandri de Vicentinis medici philosophi de calore per motum excitato, atq., de coeli influxu in sublunaria*. Verona.
Viveiros de Castro, E. 1998. "Cosmological Deixis and Amerindian Perspectivism." *Journal of the Royal Anthropological Institute,* n.s., 4:469–88.
Vogt, K. M. 2008. *Law, Reason, and the Cosmic City*. Oxford.
Volk, K., and G. D. Williams, eds. 2006. *Seeing Seneca Whole: Perspectives on Philosophy, Poetry, and Politics*. Leiden.
Voltaire 1771. *Questions sur l'Encyclopédie par des amateurs, sixième partie*. Paris.
von Leyden, W. 1985. *Aristotle on Equality and Justice*. London.

von Staden, H. 1997. "Galen and the 'Second Sophistic.'" In R. Sorabji, ed., *Aristotle and After*. London, 33–54.

———. 1998. "The Rule and the Exception: Celsus on a Scientific Conundrum." In C. Deroux, ed., *Maladie et maladies dans les textes latins antiques et médiévaux*. Brussels.

———. 2000. "Body, Soul, and Nerves: Epicurus, Herophilus, Erasistratus, the Stoics, and Galen." In J. P. Wright and P. Potter, eds., *Psyche and Soma: Physicians and Metaphysicians on the Mind-Body Problem from Antiquity to the Enlightenment*. Oxford, 79–116.

Wallace-Hadrill, A., ed. 1989. *Patronage in Ancient Society*. London.

———. 2008. *The Roman Cultural Revolution*. Cambridge.

Wallerstein, I. 1974–89. *The Modern World-System. 3 vols*. New York and San Diego.

Walton, K. 1984. "Transparent Pictures: On the Nature of Photographic Realism." *Critical Inquiry* 11:246–77.

Waters, C. K. 1998. "Causal Regularities in the Biological World of Contingent Distributions." *Biology and Philosophy* 13:5–36.

Watson, G. 1971. "The Natural Law and Stoicism." In A. A. Long, ed., *Problems in Stoicism*. London, 216–38.

Wedin, M. V. 2000. *Aristotle's Theory of Substance*. Oxford.

Weinert, F., ed. 1995. *Laws of Nature: Essays on the Philosophical, Scientific, and Historical Dimensions*. Berlin.

White, R. 2003. "The Epistemic Advantage of Prediction over Accommodation." *Mind* 112:653–83.

Whitmarsh, T. 2005. *The Second Sophistic*. Oxford.

Williams, G. 2005. "Interactions: Physics, Morality, and Narrative in Seneca, *Natural Questions* I." *Classical Philology* 100:142–65.

Wilson, L. G. 1959. "Erasistratus, Galen, and the Pneuma." *Bulletin of the History of Medicine* 33:293–314.

Wilson, M. 2007. "Rhetoric and the Younger Seneca." In Dominik and Hall 2007, 425–38.

Wood, N. 1988. *Cicero's Social and Political Thought*. Berkeley.

Woodman, T., and D. West, eds. 1984. *Poetry and Politics in the Age of Augustus*. Cambridge.

Woodward, J. 2001. "Law and Explanation in Biology: Invariance Is the Kind of Stability That Matters." *Philosophy of Science* 68:1–20.

Woolf, G. 1994. "Becoming Roman, Staying Greek: Cultural Identity and the Civilizing Process in the Roman East." *Proceedings of the Cambridge Philological Society* 40:116–43.

———. 1998. *Becoming Roman*. Cambridge.

Worrall, J. 1989. "Structural Realism: The Best of Both Worlds?" *Dialectica* 43:99–124.

———. 2006. "Theory-Confirmation and History." In Cheyne and Worrall 2006, 31–62.

Index